HANDBOOK OF GEOPHYSICAL EXPLORATION

SEISMIC EXPLORATION

Editors: Klaus Helbig and Sven Treitel

PUBLISHED VOLUMES

1984 - Mathematical Aspects of Seismology. 2nd Enlarged Edition (M. Båth and A.J. Berkhout)*
1984 - Seismic Instrumentation (M. Pieuchot) ISBN 0-08-036944-8
1984 - Seismic Inversion and Deconvolution (a) Classical Methods (E.A. Robinson)*
1985 - Vertical Seismic Profiling (a) Principles. 2nd Enlarged Edition (B.A. Hardage)*
1987 - Pattern Recognition & Image Processing (F. Aminzadeh)*
1987 - Seismic Stratigraphy (B.A. Hardage)*
1987 - Production Seismology (J.E. White and R.L. Sengbush)*
1989 - Supercomputers in Seismic Exploration (E. Eisner)*
1994 - Seismic Coal Exploration (b) In-Seam Seismics (L. Dresen and H. Rűter) *
1994 - Foundations of Anisotropy for Exploration Seismics (K. Helbig) ISBN 0-08-037224-4
1998 - Physical Properties of Rocks: Fundamentals and Principles of Petrophysics (J.H. Schőn) ISBN 0-08-041008-1
1998 - Shallow High-Resolution Reflection Seismics (J. Brouwer and K. Helbig) ISBN 0-08-043197-6
1999 - Seismic Inversion and Deconvolution (b) Dual-Sensor Technology (E.A. Robinson) ISBN 0-08-043627-7
2000 - Vertical Seismic Profiling: Principles. 3rd Updated and Revised Edition (B.A. Hardage) ISBN 0-08-043518-1
2001 - Seismic Signatures and Analysis of Reflection Data in Anisotropic Media (I. Tsvankin) ISBN 0-08-043649-8
2001 - Wave Fields in Real Media: Wave Propagation in Anisotropic, Anelastic and Porous Media (J. M. Carcione) ISBN 0-08-043929-2
2002 - Multi-Component VSP Analysis for Applied Seismic Anisotropy (C. MacBeth) ISBN 0-08-042439-2
2002 - Nuclear Magnetic Resonance. Petrophysical and Logging Applications (K.-J. Dunn, D.J. Bergman and G.A. LaTorraca) ISBN 0-08-043880-6

* *book out of print*

SEISMIC EXPLORATION

Volume 32

NUCLEAR MAGNETIC RESONANCE

PETROPHYSICAL AND LOGGING APPLICATIONS

by

K.-J. DUNN
Chevron Petroleum Technology Company
San Ramon, CA, U.S.A.

D.J. BERGMAN
Tel Aviv University, School of Physics & Astronomy
Tel Aviv, Israel

G.A. LATORRACA
Danville, CA, U.S.A.

2002
PERGAMON
An Imprint of Elsevier Science
Amsterdam – London – New York – Oxford – Paris – Shannon – Tokyo

ELSEVIER SCIENCE Ltd
The Boulevard, Langford Lane
Kidlington, Oxford OX5 1GB, UK

First edition 2002

Library of Congress Cataloging in Publication Data
A catalog record from the Library of Congress has been applied for.

British Library Cataloguing in Publication Data
Dunn, K.-J.
Nuclear magnetic resonance : petrophysical and logging applications. - (Handbook of geophysical exploration. Seismic exploration ; v. 32)
1.Seismic prospecting 2.Nuclear magnetic resonance
3.Petroleum - Prospecting
I.Title II.Bergman, D. J. III.LaTorraca, G. A.
622.1'592

ISBN 0080438806

ISBN: 0-08-043880-6
ISSN: 0950-1401 (Series)

♾ The paper used in this publication meets the requirements of ANSI/NISO Z39.48-1992 (Permanence of Paper).
Printed in The Netherlands.

To Jan Korringa and Robert J.S. Brown

Contents

Preface

The applications of nuclear magnetic resonance (NMR) to petroleum exploration and production have become more and more important in recent years. The development of the NMR logging technology and the NMR applications to core analysis and formation evaluation have been very rapid and extensive.

Part of this book was a collection of many lecture notes written over the years for Chevron in-house training purposes. As we started writing, the scope of the book gradually broadened. It covers a wide range of NMR related petrophysical measurements on cores including brief descriptions of recent applications of Magic Angle Spinning (MAS) NMR and the basics of NMR imaging of cores. In the discussion of NMR logging applications, we outline various schemes of using NMR logs to obtain necessary information for formation evaluation, such as irreducible water saturation determination, hydrocarbon typing, oil viscosity estimation, permeability prediction, etc. We discuss mainly the principles of these applications using schematic diagrams for illustration.

We believe that a unique aspect of the book is that it provides a detailed account of the basic principles of spin diffusion and relaxation in porous media. Readers who are interested in a rigorous treatment of the subject are referred to Chapter 7, where both the eigenfunction and random walk methods are described. For a less mathematical treatment, readers are referred to Chapter 4. The basic physics of nuclear magnetic resonance, as well as its relaxation behaviors, are described in Chapter 2.

The inversion of NMR data into a distribution of amplitudes associated with relaxation times provides the basic information needed to interpret the NMR measurements obtained from logging. It is an important subject. In Chapter 6, we discuss various inversion schemes as well as the procedure of regularizing the solution to be commensurate with the noise level of the raw data. There are many different schemes for implementing the inversion.

Some of them are hybrids of different methods. But the general philosophy is the same, i.e., to discard singular values of the correlation matrix which are smaller than the noise level, subject to a non-negativity constraint of the solution.

An area that we did not elaborate is the tool design and the effect of logging speed to the polarization time for the spins. Because different tools have different magnet configurations, thus different field maps, it is difficult to present a general discussion. Besides, some of these maps constitute proprietary information, which is not available to users.

During the course of writing this book, we received much assistance from our colleagues in the industry and in academia. We wish to thank Boqin Sun, Songhua Chen, Charles Flaum, Bob Brown, George Hirasaki, and Simon Stonard for critically reviewing the manuscript, pointing out mistakes, and making many valuable suggestions, and Bob Kleinberg and Chris Straley for providing some of the figure originals. In particular, we thank Boqin Sun for pointing out the need to address the issues of inhomogeneous B_0 and B_1 fields. As a result, the last section in Chapter 2 was added. He also pointed out a number of errors, which we have subsequently corrected. Finally, the encouragement and support from Chevron management are also greatly appreciated.

Unfortunately, we are sure that there remain mistakes that we have not yet discovered. We therefore welcome and will appreciate any feedback from readers so that we can make further improvements.

Keh-Jim Dunn
David J. Bergman
Gerald A. LaTorraca

Acknowledgements

The authors are indebted to the following authors and publishers/institutions for permission to reproduce figures.

Academic Press

Jackson, J.A., Burnett, L.J., and Harmon, J.F., J. Magn. Reson. **41**, 411 (1980), Figure 2.

Cooper, R.K. and Jackson, J.A., J. Magn. Reson. **41**, 400 (1980), Figure 3a.

de Swiet, T.M., Tomaselli, M., Hürlimann, M.D., and Pines, A., J. Mag. Reson., **133**, 385 (1998), Figures 2 and 3.

Latour, L.L., Kleinberg, R.L., and Sezginer, A., J. Colloid Interface Sci. **150**, 535 (1992), Figures 5 and 7.

American Physical Society

Bergman, D.J. and Dunn, K.-J., Phys. Rev. E **51**, 3401 (1995), Figures 2, 3, 6, and 8.

Bergman, D.J. and Dunn, K.-J., Phys. Rev. E**52**, 6516 (1995), Figures 4 and 5.

Bergman, D.J., Dunn, K.-J., Schwartz, L.M.,and Mitra, P.P., Phys. Rev. E **51**, 3393 (1995), Figures 2, 4, 5, and 6.

Dunn, K.-J. and Bergman, D.J., J. Chem. Phys. **102**, 3041 (1995), Figures 2, 7, 11, and 12.

Mair, R.W., Wong, G.P., Hoffmann, D., Hürlimann, M.D., Patz, S., Schwartz, L.M., and Walsworth, R.L., Phys. Rev. Lett., **83**, 3324 (1999), Figure 1.

Schwartz, L.M., Hürlimann, M.D., Dunn, K.-J., Mitra, P.P., and Bergman, D.J., Phys. Rev. E **55**, 4225 (1997), Figures 4 and 8.

Elsevier Sicence

Hürlimann, M.D., Latour, L.L., and Sotak, C.H., Magn. Reson. Imaging, **12**, 325 (1994), Figure 2.
Brown, R.J.S., Magn. Reson. Imaging, **14**, 811 (1996), Figures 1 and 5.
Kleinberg, R.L., Magnetic Resonance Imaging, **14**, 761 (1996), Figure 3.

Halliburton

Figure 1.4 of this book.

McGraw Hill

Amyx, Bass, and Whiting, *Petroleum Reservoir Engineering*, Figure 4-17.

Nature Publishing

Brown, R.J.S., Nature, **189**, No.4762, 387 (1961), Figure 1.

Oxford University Press

Paul T. Callaghan, *Principles of Nuclear Magnetic Resonance Microscopy*, Figure 3.14.

Schlumberger

Figure 1.5 of this book.

Society of Core Analysts

LaTorraca, G.A., Dunn, K-J., and Brown, R.J.S., International Symposium of the Society of Core Analysts, Paper 9312 (1993), Figure 3b.
Wilson, D.M. and LaTorraca, G.A., Paper 9923, International Symposium of the Society of Core Analysts, Golden, CO (1999), Figures 3, 4, and 5.

Society of Petroleum Engineers

Brown, R.J.S. and Fatt, I., "Measurements of fractional wettability of oilfield rocks by the nuclear magnetic relaxation method," Petroleum Transaction, AIME, **207**, 262 (1956), Figure 2.

Lo, S-W., Hirasaki, G.J., House, W.V., and Kobayashi, R., "Correlations of NMR relaxation time with viscosity, diffusivity, and gas/oil ratio of methane/hydrocarbon mixture," SPE 63217 (2000), Figures 1,3,8,9, and 13.

Kleinberg, R.L., Straley, C., Kenyon, W.E., Akkurt, R., and Farooqui, S.A., "Nuclear magnetic resonance of rocks: T_1 vs. T_2," SPE 26470 (1993), Figures 4 and 5.

Matteson, A., Tomanic, J.P., Herron, M.M., Allen, D.F., and Kenyon, W.E., "NMR relaxation of clay-brine mixtures," SPE 49008 (1998), Figures 3 and 4.

Society of Professional Well Log Analysts

Chang, D., Vinegar, H.J., Morris, C.E., and Straley, C., SPWLA 35th Annual Symposium, Paper A (1994), Figure 10.

Chitale, D.V., Gardner, J., and Sigal, R., SPWLA 41st Annual Logging Symposium, Paper X, Dallas, TX, June 4-7 (2000), Figure 4.

Coolidge, J.E., Ritchie, I., Gamson, B.W., and Shane, L., Paper 11, Third Annual Logging Symposium of Society of Professional Well Log Analysts, Houston, TX (1962), Figure 11.

Kleinberg, R.L. and Vinegar, H.J., The Log Analyst, Nov.-Dec., 20 (1996), Figures 4, 9, 13, and 14.

LaTorraca, G.A., Stonard, S.W., Webber, P.R., Carlson, R.M., and Dunn, K.J., SPWLA 40th Annual Logging Symposium, Paper PPP, Oslo, Norway, May 30-June 3 (1999), Figures 2, 3, 4, and 5.

Lo, S-W., Hirasaki, G.J., Kobayashi, R., and House, W.V., The Log Analyst, Nov.-Dec. 43 (1998), Figure 2.

Morriss, C.E., Freedman, R., Straley, C., Johnston, M., Vinegar, H.J., and Tutunjian, P.N., SPWLA 35th Annual Logging Symposium, Paper C, June 19-22 (1994), Figures 4, 5, and 7.

Ramakrishnan, T.S., Fordham, E.J., Venkataramanan, L., Flaum, M., and Schwartz, L.M., SPWLA 40th Annual Logging Symposium, Paper MMM (1999), Figure 11.

Sezginer, A., Cao Minh, C., Heaton, N., Herron, M., Freedman, R., and Van Dort, G., SPWLA 40th Annual Logging Symposium, Paper NNN, Oslo, Norway, May 30-June 3 (1999), Figure 1.

Straley, C., Rossini, D., Vinegar, H.J., Tutunjian, P.N., and Morriss, C.E., The Log Analyst, **38**, 84 (1997), Figures 1, 3, 7, 8, 9, and 10.

Zhang, G.Q., Lo, S.-W., Huang, C.C., Hirasaki, G.J., Kobayashi, R., and House, W.V., SPWLA 39th Annual Logging Symposium, Paper FF, Keystone, CO., May 26-29 (1998), Figure 4.

Fundamental constants

Constant	Symbol	Value SI	CGS
Speed of light	c	2.997925×10^{8} m s^{-1}	$\times10^{10}$ cm s^{-1}
Electron charge	e	1.602192×10^{-19} C 4.80325	$\times10^{-20}$ e.m.u. $\times10^{-10}$ e.s.u.
Electron rest mass	m_e	9.10956×10^{-31} kg	$\times10^{-28}$ gm
Proton rest mass	M_p	1.67261×10^{-27} kg	$\times10^{-24}$ gm
Neutron rest mass	M_n	1.6747×10^{-27} kg	$\times10^{-24}$ gm
Planck constant	h $\hbar$	6.62620×10^{-34} J s 1.054592×10^{-34} J s	$\times10^{-27}$ erg s $\times10^{-27}$ erg s
Charge-to-mass ratio for electron	e/m_e	1.758803×10^{11} C kg^{-1}	$\times10^{7}$ e.m.u. g^{-1}
Permeability constant	μ_0	$4\pi\times10^{-7}$kg m s^{-2} A^{-2} 1	– $\times1$ e.m.u.
Molar volume at STP	V_0	22.4×10^{-3} m^{3} mole^{-1}	$\times10^{3}$ cm^{3} mole^{-1}
Avogadro's number	N_0	6.022169×10^{26} kmole^{-1}	$\times10^{23}$ mole^{-1}
Gas constant	R_0	8.31434×10^{3} J kmole^{-1}K^{-1}	$\times10^{7}$ erg mole^{-1}K^{-1}
Boltzmann's constant	k_B	1.380622×10^{-23} J K^{-1}	$\times10^{-16}$ erg K^{-1}
Gravitational constant	G	6.6732×10^{-11} N m^{2} kg^{-2}	$\times10^{-6}$ dyne cm^{2} g^{-2}

Bohr magneton $e\hbar/2m_ec$	μ_B	9.27410×10^{-24} J T^{-1}	$\times10^{-21}$ erg Gauss^{-1}
nuclear magneton $e\hbar/2M_pc$	μ_N	5.05095×10^{-27} J T^{-1}	$\times10^{-24}$ erg Gauss^{-1}
gyromagnetic ratio of protons in H_2O	γ_p	2.675197×10^{8} rad s^{-1} T^{-1}	$\times10^{4}$ rad s^{-1} Gauss^{-1}
	$\gamma_p/2\pi$	4.257707×10^{7} Hz T^{-1}	$\times10^{3}$ Hz Gauss^{-1}

$\mu_x = g_x\mu_N\sqrt{I(I+1)} = \gamma_x\hbar\sqrt{I(I+1)}$
μ_x/μ_N for ^{1}H = 4.83724

The magnetic field for 100 MHz for proton is
H = 2.34868 Tesla
1 Tesla = 10^4 Gauss

1 eV = 1.60219×10^{-12} erg
1 eV = 23.05 kcal mole^{-1}
1 bar = 14.5038 psi = 10^6 dynes/cm^2
1 atm = 1.01325 bar
1 GPa = 10 kbar
1 Pa = 1 Newton/m^2 = 10 dynes/cm^2

1 cP = 10^{-2} gm cm^{-1} s^{-1} = 1 mPa s
1 cSt = 10^{-2} cm^2/s = 1 mm^2/s

1 barrel = 42 gal
1 gal = 3.785412×10^{-3} m^3
1 Darcy = 9.869×10^{-9} cm^2

Chapter 1

Historical developments of NMR logging

1.1 Introduction

The application of nuclear magnetic resonance (NMR) to petrophysics for studying the properties of water and/or oil saturated rocks dates back to the early 1950's. By the 1960's, nuclear magnetic resonance logging tools using the earth's magnetic field were built for oil exploration. However, these tools did not gain wide popularity due to various limitations. It was not until the early 1990's that pulsed NMR type logging tools[1] were introduced and gained a wide acceptance by the petroleum industry. Since then, numerous applications have evolved along with significant tool improvements. Users can now extract information, not only on the saturating fluids in the rocks, but also on the pore size distributions, thus the pore fabric, of different rock types.

The information we obtain from the NMR properties of the fluid-saturated rocks is primarily based on the nuclear spin relaxation. NMR spectroscopy used in the identification of chemical states is seldom used in the studies of rocks. While there are numerous books that deal with NMR spectroscopy, very few provide a thorough account of NMR relaxation in fluid-saturated porous media and its applications to petroleum exploration. It is the primary objective of this book to fill that gap.

[1]Logging tools are instruments lowered into a drilled borehole in the earth formation to make physical measurements for hydrocarbon or mineral prospecting.

1.2 Nuclear magnetic resonance

The phenomenon of nuclear magnetic resonance was first observed independently in 1946 by scientists at Stanford (Bloch [22, 23]) and Harvard (Purcell et al. [181], Bloembergen et al. [24]) Universities. The subject quickly grew into a sophisticated technique with wide applications in physics and chemistry. Many of these applications have to do with investigations of microscopic interactions in materials at the molecular and atomic levels. Intricate interactions often manifest themselves in shifts of nuclear magnetic resonance frequency. Thus, NMR becomes a powerful tool for the identification of chemical states in matter, a technique commonly known as NMR spectroscopy.

NMR also found applications in biomedicine, food, and petroleum industries. In the biomedical field, where NMR is used as an imaging tool for medical diagnosis, there has been extensive development in the last decade. The application of NMR for petroleum exploration started as early as 1950's. Scientists at Mobil, Exxon, Shell, Texaco, and Chevron were among the early pioneers in the use of NMR technology [226]. Many of the early applications of NMR spectra were investigations of the composition of petroleum and petroleum fractions. Chevron, for some time, maintained an extensive program on the application of NMR to petrophysical studies of rocks and fluids and in collaboration with Byron-Jackson and Borg-Warner, and in consultation with Varian Associates, developed a well logging tool for formation evaluation[2].

The NMR signal from hydrogen nuclei is proportional to the population of hydrogen atoms in the sample, and the signal relaxation rate is proportional to the viscosity of the fluid. Therefore, it was natural to consider using these NMR properties for oil exploration. However, early investigators found that the nuclear magnetic relaxation of water in rocks is much faster than that in the bulk phase. The increase of the relaxation rate is primarily due to the surface relaxivity and is related to the surface to volume ratio of the pore space. This discovery led to the demonstration of the connection between the NMR properties of porous media and their permeability. The early research work in Chevron [194, 212, 37, 216] as well as others (e.g. Schlumberger, Mobil, Texaco, Borg-Warner, and Varian [123]) not only resulted in a number

[2]Formation evaluation is a branch of earth sciences for evaluating petrophysical properties of the earth formation from measurements of many logging tools.

of publications illustrating the important applications of NMR technology for oil exploration, but also led to the development of an experimental logging tool [38] and to commercial logging services in the 1960's by Pan Geo Atlas Corp. (PGAC) [66] and by Schlumberger [100]. These efforts finally led to the extensive developments of the modern day NMR logging technology (Miller et al. [163], Prammer et al. [176], Kleinberg et al. [124], Freedman et al. [87]).

The commercial Nuclear Magnetism Logging (NML) tools, when and where available, operated by the free precession in the earth's magnetic field until 1994, when the first modern NMR logging tools operating at higher magnetic fields became commercially available after several years of field testing. The modern day logging tools carry their own magnets, and operate at higher Larmor frequencies. Each service company has its own commercial name (Halliburton/Numar's is called Magnetic Resonance Imaging Log (MRIL), Schlumberger's is called Combinable Magnetic Resonance tool (CMR)).

1.3 NMR logging tool developments

The initial idea for a NMR logging tool came from Russell Varian [36], and subsequent feasibility studies were conducted at Varian Associates from 1953 to 1955. At that time, a number of oil companies as well as Schlumberger started to do important work on NMR in fluid-saturated porous media. The initial thought was that one could use the Nuclear Magnetism Logging tool to measure hydrogen in both oil and water in porous rocks, and they would be distinguishable by different relaxation times, with water usually having the longer relaxation time. Since water and oil have about the same hydrogen density, one would expect to obtain a measurement of the total porosity of the rock as well.

Subsequent developments were largely unexpected. Nobody anticipated that the surface effects on relaxation of water in the pore space would be so significant. Water in the pores of most rocks was found to have a wide range of relaxation times due to surface relaxivity. The oil relaxation time in the rocks, on the other hand, was found *usually* to have the same relaxation time as that in the bulk phase[3]. Because the oil relaxation time is likely to be the

[3]Note, for oil wet rocks, oil also behaves similarly to water, having a wide range of relaxation times. The fact that oil relaxation time in rocks is usually the same as that of the bulk phase is because most rocks are water wet. (See discussion in Chapters 3 and 4)

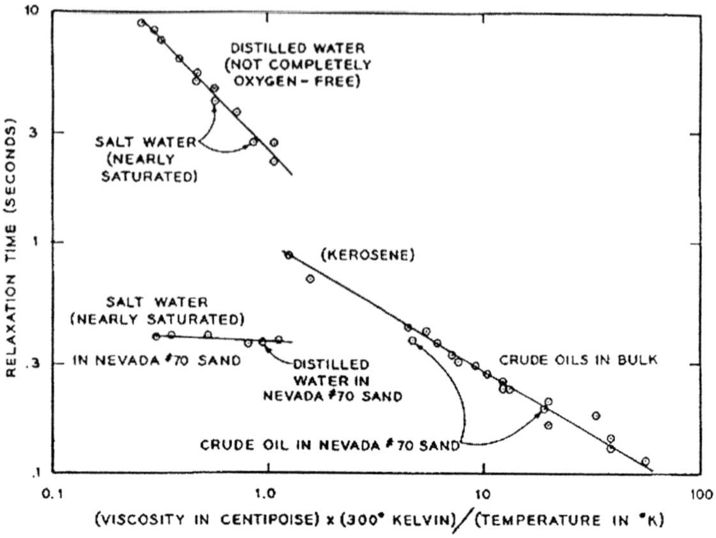

Figure 1.1: Relaxation times in bulk fluids as well as those in the porous media. (From Brown [36].)

same as for some of the water in the rock, it becomes a nontrivial matter to differentiate oil from water simply using their different relaxation times.

Figure 1.1 shows the understanding of relaxation times of various fluids in bulk phase and in porous media at the end of 1955 [36]. Relaxation times were plotted against viscosity divided by absolute temperature. The bulk water and various bulk oils show longer relaxation times for lower viscosity. An important observation is that putting oil in the pores of a sand has very little effect on its relaxation time, whereas putting water in the pores of a sand greatly reduces relaxation times and almost eliminates the dependency of relaxation time on viscosity or temperature. This behavior is found to be due to the surface relaxation effect in the pore space, with water being in contact with the pore surface, whereas the oils usually not.

In addition to the complexity of not being able to differentiate water and oil in the rocks, the early development of NML ran into another obstacle. The tool was designed based on the observation of proton free precession in the earth's magnetic field at a frequency of about 2 kHz. The coil (shown schematically in Figure 1.2) was 1.75 m long and 15 cm wide. In operation, the coil was energized with a strong dc current to produce a field for a few

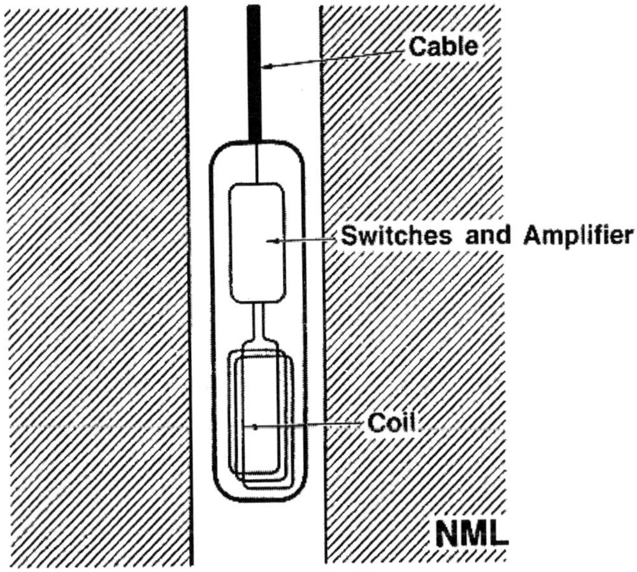

Figure 1.2: Early Chevron design of a nuclear magnetic logging tool. (From Brown [36].)

seconds while a nuclear magnetization was built up. Then the current was abruptly turned off, and the same coil was used to receive the signal from the proton free precession in the earth's magnetic field. The instrument dead-time was about 30 ms (later reduced to 21 ms). As a result, part of the water in the pore space with relaxation times less than the dead time was not measurable. Thus the NML tool would not be able to distinguish the water and oil in pore space; it would not even be able to measure the total porosity.

With these new understandings, the development of the NML was continued with different objectives. Petrophysical studies in the laboratory have shown that the relaxation times of water are closely related to surface to volume ratio of the pore space. Large surface to volume ratio has small permeabilities, and vice versa. Thus, the NMR measurements can be used to predict permeability. Fortuitously, the dead time was quite close to the T_2 cutoff established for irreducible water saturation. The fluid that was measurable by NML was called the "Free Fluid". This term continues to be used now to describe the portion of the fluid in the pore space that is movable, as "Free Fluid Index" (or FFI).

An NML tool was finally developed based on the earth's field concept (Brown and Gamson [38]) and was used in field measurements [100]. But the tool did not gain wide acceptance due to several limitations. A major problem was that the tool measures the signals from the water and oil in the formation, which are of interest to us, as well as the signals from the water-filled borehole, which are not. Thus, to kill the signals from the borehole, powdered magnetite (magnetic iron oxide) was added to the borehole fluid. This made locally inhomogeneous fields that shortened signal lifetimes so much that signals were not measurable with a 30 ms dead time instrument. However, this procedure is frequently undesirable due to other operational constraints, and it also excludes applications in deep wells. The sonde housing must be non-magnetic, which limits its use in hostile environment. Besides, the tool sensitivity depends on the tool orientation with respect to the earth's magnetic field. This creates complications in the signal calibration.

To circumvent the problems associated with using the earth's magnetic field, people started thinking of using permanent magnets to generate a homogeneous field at an external location outside the probe, the so-called remote inside-out concept. Cook [65] and Cooper and Jackson [67] independently suggested a method for producing a toroidal region of homogeneous magnetic field outside the NMR probe. In the design by Jackson and his colleagues [47, 112], two cylindrical magnets are placed in a tool housing with their north poles opposing each other. These magnets generate a static field which is radially symmetric. There exists a toroidal region on the horizontal mid plane perpendicular to the central axis where the radial component of the magnetic field H_r can be made homogeneous to any chosen degree over prescribed regions.

Figure 1.3(a) shows the magnet configuration for the production of a toroidal region of homogeneous radial magnetic field. Figure 1.3(b) shows the three dimensional plot of the radial component of the magnetic field as a function of the radial distance and the separation between the opposing magnets. Notice that for a fixed separation, i.e., h/a (where h is the distance between the opposing magnets, a is the radius of the cylindrical magnets, and r is the radial distance from the origin at the mid plane between the magnets), the radial component of the magnetic field, H_r, goes through a maximum. In the neighborhood of this maximum, the field can be made uniform over a certain region, to any desired accuracy, by increasing the separation of the opposing magnets, h/a. The inset of Figure 1.3(a) indicates a coil, when

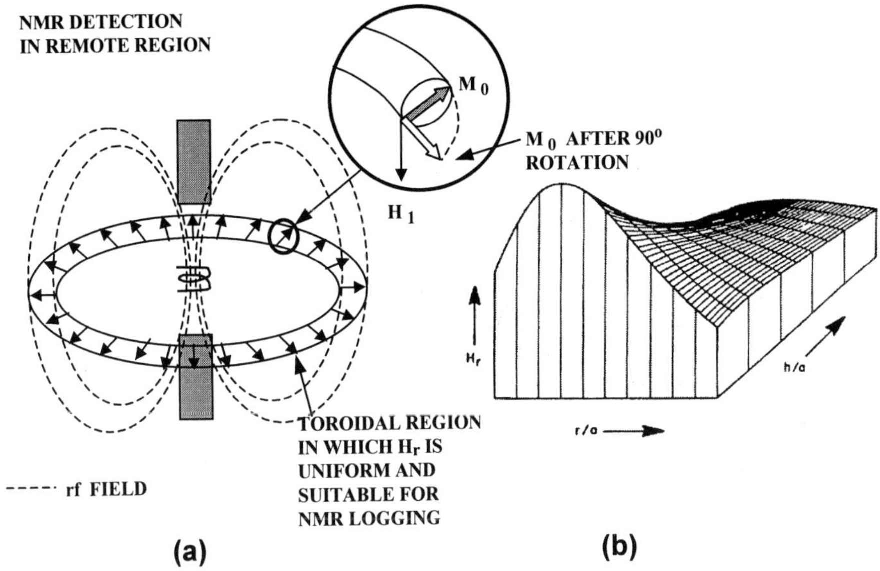

Figure 1.3: (a) Toroidal region of homogeneous field of two opposing cylindrical magnets (From Jackson et al. [112]). (b) Three dimensional plot of the radial field H_r versus magnet separation h/a and radial distance r/a. (From Cooper and Jackson [67].)

placed in the middle of the opposing magnets, that can generate a 90° pulse in the sensitive region. Following the application of the radio frequency (rf) pulse, the magnetization precesses in a plane perpendicular to the magnetic field H_r. The coupling of the precessing magnetization and the coil allows the NMR signal to be detected. For a detailed understanding of how the tool works, the readers are referred to Chapter 2.

Later on, Clow et al. [62] built a NMR logging tool based on Jackson's and his colleagues' design. They placed in the copper solenoid a core of a high permeability ferrimagnetic material, such as manganese-zinc ferrite, thus greatly increasing the Q factor and improving the signal to noise ratio.

A basic difficulty with the Jackson design is that the location and field strength, of the toroidal region of highly homogeneous field, tend to vary over time during operation, due to changes in local earth's magnetic field, formation temperature, and mechanical parameters of the field producing permanent magnets. When the rf coil is tuned to a fixed frequency, it lacks the flexibility to resonantly match the varying field condition.

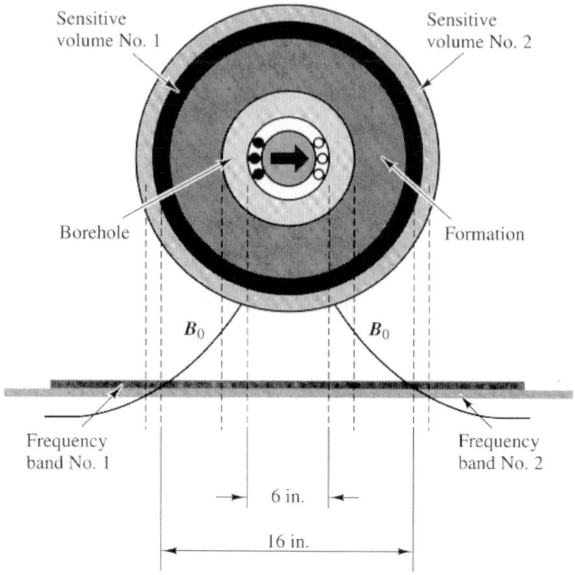

Figure 1.4: The cross sectional view of Numar/Halliburton's Magnetic Resonance Imaging Log (MRIL) tool based on Shtrikman-Taicher's design. The tool is centered in the borehole, and the probed region is a thin cylindrical shell. (Courtesy of Numar/Halliburton Corp.)

Shtrikman and Taicher [200] came up with a different design in which a long cylindrical magnet is magnetized transversely to its long axis, producing a static magnetic dipole field in the earth formation which varies as $1/r^2$. The rf coil is wound lengthwise on the cylindrical magnet in the plane containing the magnetization direction and the longitudinal axis (Figure 1.4). The resonant condition is satisfied on a thin cylindrical shell coaxial with the magnet. The probed region of this thin cylindrical shell has a length roughly equal to the long dimension of the coil along the axis.

In the Shtrikman-Taicher design, the magnetic field tapers off as $1/r^2$. The tuned rf coil will always select a thin cylindrical shell whose field strength matches the resonance frequency, thus alleviating the resonance matching problem faced by the Jackson design. The Shtrikman-Taicher design was later developed into the Numar/Halliburton's Magnetic Resonance Imaging Log (MRIL) tool. One of the weaknesses of this design is the smallness of the probed dimension in the radial direction, approximately 1 to 2 mm thick, and the field is not homogeneous, having a well-defined field gradient. Such a

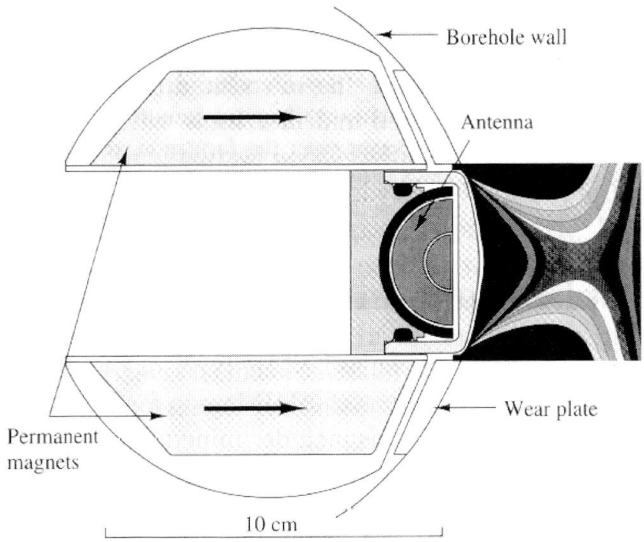

Figure 1.5: The cross sectional view of Schlumberger's Combinable Magnetic Resonance (CMR) tool. The sonde is pressed against the borehole wall. The probe region has a saddle-point configuration. Its center contains an area with relatively homogeneous magnetic field. (Courtesy of Schlumberger Ltd.)

weakness is compensated by the long dimension of the cylindrical shell, which results in a large probed volume, and a consequent adequate signal to noise ratio. The presence of a field gradient, initially thought to be a drawback, turned out to be a desirable feature for the application of hydrocarbon typing (See discussion in NMR log applications in Chapter 5).

In contrast to the centralized design of Shtrikman and Taicher, scientists in Schlumberger (Kleinberg et al. [124]) designed a pad-type tool. A cross-sectional view of their design is shown in Figure 1.5. The sonde consists of three slab magnets magnetized in the same direction, pressed against the borehole wall, resulting a homogeneous field in a region about 0.75 in inside the formation with a width about an inch in the radial direction and about 6 in along the the borehole wall. The antenna is filled with ferrite to increase its sensitivity. Such a design has a high depth resolution and a low sensitivity to the borehole fluid salinity. It was later developed into Schlumberger's Combinable Magnetic Resonance tool (CMR).

The NMR logging tools currently on the market, Halliburton/Numar's

which give rise to nuclear magnetism. For example, the hydrogen nucleus, ^{1}H, has a spin of $\frac{1}{2}$, which will lead to nuclear magnetic resonance. Other nuclei, such as ^{11}B, ^{13}C, ^{15}N, ^{19}F, ^{23}Na, etc., all have spin angular momenta and magnetic moments. In NMR logging, we are interested only in proton (^{1}H) NMR.

Electrons also have spin angular momentum and an associated magnetic moment. In fact, because of their small mass, their magnetic moment is much stronger than that of the hydrogen nucleus. Similar to the nuclear magnetic resonance of protons, electrons also have magnetic resonance, which is called *electron spin resonance* (ESR). Generally speaking, the nuclear properties such as spin and magnetic moment are, to a very high degree of approximation, unaffected by the surroundings, whereas electrons are strongly dependent on the surroundings due to the larger interaction energy.

There are other magnetic moments associated with the motion of electrons. When electrons orbit a nucleus, their orbital motion produces a magnetic moment. If an external magnetic field is applied to the system, these electrons will orbit the nucleus to produce a magnetic field in the direction opposite to the applied field (Lenz law). This leads to a negative magnetization, i.e., $M = \chi H$, where M is the magnetization, H is the applied field, and χ is the magnetic susceptibility and is negative, typically about -10^{-6} emu. Such atoms are called *diamagnetic.*

Electrons have their own spin and magnetic moment. When they are paired together with one spin up and one spin down, the atom they reside on does not respond to an applied magnetic field. When they are unpaired, the net magnetic moment tends to align with the external applied field. If this net increase of magnetic moment more than compensates the diamagnetism, the atom will have a positive susceptibility, typically about 10^{-6} emu or much higher. Such atoms are called *paramagnetic.*

For a paramagnetic material, as the temperature is lowered, the interaction between neighboring paramagnetic atoms will become stronger due to weaker thermal agitations. When the temperature is lowered below a threshold temperature (commonly known as the Curie temperature), the paramagnetic atoms tend to align their magnetic moments with one another and form *magnetic domains.* A material in this state is called *ferromagnetic.* When it is placed in a magnetic field, the magnetic domains with spin direction along the field direction will grow at the expense of other domains, and become permanently magnetized even after the field is removed. Paramagnetic atoms (above the Curie temperature) also align with the field but do not retain their alignment after the field is removed.

Many atoms are paramagnetic to some degree, but there is a wide range; the magnetism produced by the stronger paramagnetic atoms is 1,000 times as strong as the nuclear magnetism of hydrogen nuclei. In the face of the much larger paramagnetism due to electrons, nuclear magnetism can only be studied because it is highly resonant at a specific frequency. Manganese, gadolinium, and iron are strongly paramagnetic. Their presence in small amounts (ppm) can greatly reduce the T_1 and T_2 time constants of rocks containing these atoms in the pore fluids and on the grain surfaces.

2.1.2 Precession of spins

The following analysis applies to magnetic nuclei which possess both a magnetic moment and angular momentum. Let us assume the hydrogen nucleus, a proton, has a spin angular momentum $\mathbf{J}$ and an associated magnetic moment μ. Since they are parallel to each other, we can write

$$\mu = \gamma \mathbf{J}, \tag{2.1}$$

where γ is called gyromagnetic (or magnetogyric) ratio. In principle, one can calculate both μ and $\mathbf{J}$, and thus γ. For our purpose, we only need to know that γ is a constant, which has different values for different nuclei. For protons, $\gamma_p = 2.675197 \times 10^8$ rad s^{-1} Tesla^{-1} (or $\times 10^4$ rad s^{-1}Gauss^{-1}), or $\gamma_p/2\pi = 4.257707 \times 10^7$ Hz/Tesla (or $\times 10^3$ Hz/Gauss).

When a proton is placed in an external magnetic field $\mathbf{H}$, the magnetic field will produce a torque on the magnetic moment μ of amount $\mu \times \mathbf{H}$. If this magnetic moment acted like an ordinary bar magnet, it would simply align along the direction of $\mathbf{H}$ under the action of the torque. If the magnet also possessed angular momentum, as in the case for hydrogen nucleus, then the situation would be modified. Because hydrogen nucleus possesses angular momentum, it instead acts like a gyropscope and precesses about the external field $\mathbf{H}$.

The equation of motion for this precession is found by equating the torque with the rate of change of angular momentum $\mathbf{J}$ for the nucleus

$$\frac{d\mathbf{J}}{dt} = \mu \times \mathbf{H}. \tag{2.2}$$

Since $\mu = \gamma \mathbf{J}$, we get

$$\frac{d\mu}{dt} = \mu \times (\gamma \mathbf{H}). \tag{2.3}$$

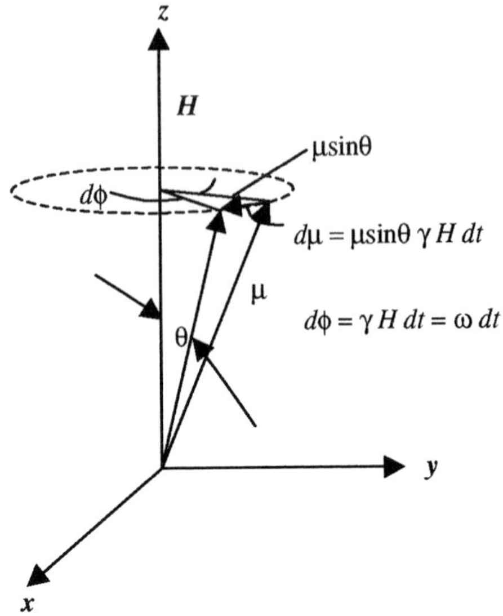

Figure 2.1: A spin of magnetic moment μ is precessing about the applied magnetic field H (along z-axis) with a Larmor frequency $f = \omega/2\pi = \gamma H/2\pi$.

This equation tells us that at any instant the changes in μ are perpendicular to both μ and $\mathbf{H}$. Referring to Figure 2.1 and considering the tail of the vector as fixed, the tip of the vector is therefore moving in the direction which is perpendicular to the plane formed by μ and $\mathbf{H}$. Hence the angle θ between μ and $\mathbf{H}$ does not change, and the motion of μ forms a cone about $\mathbf{H}$. Since in time dt, the tip of the vector μ moves an angle $(\gamma H)dt$, the vector μ is precessing about $\mathbf{H}$ at a frequency of $\gamma H/2\pi$. This frequency, $f = \omega/2\pi = \gamma H/2\pi$, is called the *Larmor frequency*.

For example, a hydrogen nucleus precesses at the rate of 42.6 million cycles per second (42.6 MHz) for a field of 1 Tesla (10,000 Gauss). In contrast, the earth's field is approximately 0.5 gauss at temperate latitudes. The Larmor frequency is important because it is the resonance frequency of the magnetic nuclei. Magnetic nuclei can absorb/emit energy only at their Larmor frequencies.

2.2 Relaxation times

2.2.1 Longitudinal relaxation

Now let us consider a system containing N_0 identical nuclei, each having a magnetic moment μ. It is assumed that the magnetic interactions of the nuclei with each other are small compared to their interaction with the applied field H in the z direction. Hence each nuclear moment has quantized energy levels given by

$$E_m = -\mu \cdot \mathbf{H} = -\gamma\hbar\mathbf{I} \cdot \mathbf{H} = -\gamma\hbar H m, \tag{2.4}$$

where $\hbar$ is Planck's constant h divided by 2π, $\hbar\mathbf{I} = \mathbf{J}$, and $\mathbf{I}$ is a dimensionless angular momentum operator. $\mathbf{I}^2$ and I_z commute with each other and have eigenvalues $I(I+1)$ and m, respectively, with I either an integer or a half integer (i.e., $2I$ is always an integer) and m varying from $-I$ to $+I$ in steps of 1.

At thermal equilibrium, the system can be described by Boltzmann statistics. There are N_m nuclei occupying the energy level E_m with N_m given by

$$N_m = \frac{N_0 e^{-E_m/kT}}{\sum_{m'} e^{-E_{m'}/kT}} = \frac{N_0 e^{\gamma\hbar H m/kT}}{\sum_{m'} e^{\gamma\hbar H m'/kT}}, \tag{2.5}$$

where the summation extends from $-I$ to $+I$, with $m = +I$ being the lowest energy state. The occupation numbers N_m satisfy the condition

$$N_0 = \sum_m N_m, \tag{2.6}$$

and the steady nuclear polarization along the z axis is given by

$$M_0 = \sum_m N_m \mu_z = \gamma\hbar \sum_m m N_m, \tag{2.7}$$

where M_0 is the magnetization of the system.

For simplicity, we consider the case of hydrogen nuclei. I equals $\frac{1}{2}$, and there are only two quantized states, one with spin up, $m = +\frac{1}{2}$, parallel to the applied field; the other one with spin down, $m = -\frac{1}{2}$, antiparallel to the applied field. At thermal equilibrium, we have

$$\frac{N_-}{N_+} = e^{-\mu H/kT} = e^{-\gamma\hbar H/kT}, \tag{2.8}$$

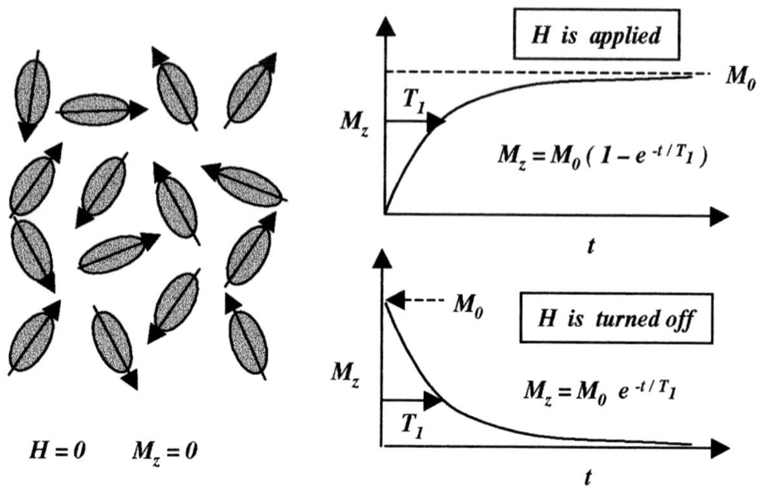

Figure 2.2: When there is no magnetic field, the proton spins are randomly oriented, resulting in zero magnetization. As the field H is applied in z-direction, the magnetization of the system, M_z, gradually builds up until it reaches equilibrium value, M_0. The process is reversed when H is set to zero. T_1 is defined as the characteristic time when the increasing M_z reaches $M_0(1 - e^{-1})$ or the decreasing M_z reaches $M_0 e^{-1}$.

where the lower energy state population N_+ is slightly larger than the higher energy state population N_-. Note that for a field which leads to a Larmor frequency of 1 MHz for protons, the magnetic energy μH is of the order of 6.6×10^{-21} erg. This is to be compared with the thermal energy kT at room temperature of about 4×10^{-14} erg. Hence, $e^{-\mu H/kT} \approx 1 - 1.7 \times 10^{-7}$, i.e. N_- is only slightly less than N_+ by about 2 parts per 10 million. This means the net magnetization is extremely small due to the relatively large thermal agitations of the nuclei.

When the applied field $\mathbf{H}$ is zero, the ratio in Eq. (2.8) is one, which means $N_- = N_+$, and the net magnetization is zero. When a nonzero $\mathbf{H}$ is suddenly applied in the z direction, some higher energy nuclei undergo transitions to the lower energy state by emitting energy $\hbar\omega (= \mu H)$ to the surrounding molecules. With time, the population of N_+ gradually increases while that of N_- decreases to values described by a Boltzmann distribution, i.e., Eq. (2.8). The magnetization of the system M_z gradually reaches the thermal equilibrium value M_0. The process is reversed when $\mathbf{H}$ is set to zero, the lower energy nuclei undergo transitions to the higher energy state by absorbing energy $\hbar\omega$ from the surrounding molecules and eventually M_z becomes zero. Figure 2.2 shows the relaxation behavior of the magnetization of a system of protons as the applied field H is turned on and off along the z axis.

The environment of a nucleus with surrounding molecules is often summarily described as a lattice. The thermal motions of the lattice set up fluctuating electric and magnetic fields at the nucleus. The interaction of the nuclear moment with these fields helps stimulate the transitions between the magnetic energy levels by emitting or absorbing energy to or from the surroundings. This kind of process, called *spin-lattice relaxation*, eventually leads to thermal equilibrium. Bloch [23] introduced the assumption that the change of M_z with time can be described by the following equations

$$\begin{aligned} \frac{dM_z}{dt} &= \frac{M_0 - M_z}{T_1} \\ M_z &= M_0(1 - e^{-t/T_1}) \end{aligned} \tag{2.9}$$

where T_1 is called the *longitudinal*, or *spin-lattice relaxation time*. The time constant T_1 is a joint property of the spin system and its environment. Its value reflects how effectively the magnetic energy of the spin system is transferred to or from its surroundings. A large value of T_1 corresponds to weak coupling and a slow approach to equilibrium, whereas a small value of T_1 indicates strong coupling and a rapid approach to equilibrium.

2.2.2 Transverse relaxation

In the previous section, we assumed that an external magnetic field $\mathbf{H}$ was applied along the z direction, which caused the spins to align themselves along that direction. As the thermal equilibrium was reached, a net magnetization M_0 was formed along the z axis. As shown in Figure 2.3, we now place a radio frequency (rf) coil around the spin system with the coil axis perpendicular to the z axis, say in the x direction. If we apply current through the coil, with a carrier frequency equal to the Larmor frequency, it generates a linearly polarized, oscillatory transverse magnetic field $2H_1\mathbf{e}_x \cos\omega t$, where H_1 is the rf field strength, $\mathbf{e}_x$ is the unit vector along x axis, and $\omega = \gamma H$

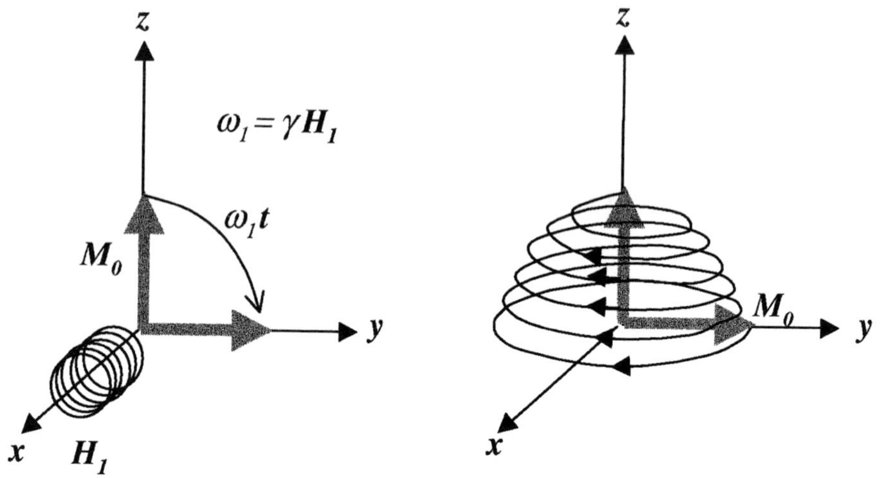

Figure 2.3: When a rf pulse is applied through a coil placed along the x-axis, a magnetic field H_1 is generated, causing M_0 to tilt into y-axis. As the pulse is turned off, the magnetization, M_0, precesses about z-axis, and gradually returns to z-direction. The projection of M_0 on the xy plane is M_x and M_y.

is the Larmor angular frequency. It is convenient to represent the linearly polarized oscillatory field as a sum of two circularly polarized fields:

$$2H_1\mathbf{e}_x \cos\omega t = \begin{cases} H_1(\mathbf{e}_x \cos\omega t + \mathbf{e}_y \sin\omega t) \\ H_1(\mathbf{e}_x \cos\omega t - \mathbf{e}_y \sin\omega t) \end{cases} \tag{2.10}$$

rotating at opposite directions about z axis, one clockwise, and the other counter-clockwise. One of them will rotate in the same direction as the nuclear spin precession and will be responsible for the resonance phenomenon (clockwise in the present case for proton resonance). The other one can be ignored if $H_1 \ll H$.

In a rotating frame about z axis with an angular velocity ω, the rotating field $\mathbf{H_1}$ is stationary and along x axis. When we apply a pulse through the rf coil, it causes the magnetization M_0 to precess about the x axis in the rotating frame. The pulse width can be adjusted such that the magnetization M_0 rotates by exactly 90°; then the pulse terminated and $\mathbf{H_1}$=0. Now the only field left is the externally applied fixed field $\mathbf{H}$ in the z-direction. The magnetization M_0 will then precess about $\mathbf{H}$ in the laboratory frame as shown in Figure 2.3, initially in the xy plane, gradually moving up, and eventually returning to the z direction. The decays of M_x and M_y in the xy plane of the rotating frame can be described by the following Bloch equations

$$\frac{dM_x}{dt} = -\frac{M_x}{T_2}, \quad \frac{dM_y}{dt} = -\frac{M_y}{T_2} \tag{2.11}$$

where T_2 is called the *transverse*, or *spin-spin relaxation time.*

Transverse reminds us that the T_2 relaxation process occurs in the xy plane, whereas the T_1 refers to the decay time of magnetization in the z direction. Any process that causes the loss of transverse magnetization, including the return to the z axis, contributes to T_2. The first thing that comes into consideration is the inhomogeneity of the static field $\mathbf{H}$. Recall that the magnetization vector M_0, when precessing in the xy plane about $\mathbf{H}$, is actually composed of all spins of the system, all of them precessing about the z axis. In general, the static field is not perfectly uniform over the whole system. As a result, spins are precessing at slightly different Larmor frequencies, and the spin precessions quickly get out of phase. For example, a one part per million variation of the static field between nearby regions will cause the nuclei in the two regions to be 180° out of phase in about 10 ms for a magnetic field of 1 Tesla. The decay of magnetization in the xy plane is frequently dominated by the field inhomogeneity effect, and the transverse relaxation time is generally referred to as T_2^*. It turns out that it is possible to experimentally separate the loss of magnetization due to macroscopic field inhomogeneity (e.g. CPMG method which we shall discuss later) from that due to other causes. The symbol T_2 is usually reserved for the latter.

What are these other causes that would contribute to the T_2 relaxation besides nonuniformity of the applied static field $\mathbf{H}$? In solids, the nuclei are not free to move around. Hence, no matter how uniform the applied field is, the local magnetic fields due to neighboring nuclei in the material can cause T_2 to be very short. In contrast, the nuclei in liquids move so fast that they average out the varying local fields so quickly that the only cause for transverse relaxation is the returning of the magnetization to the z axis. Thus frequently T_2 equals T_1 in liquids. However, T_2 can never be longer than T_1.

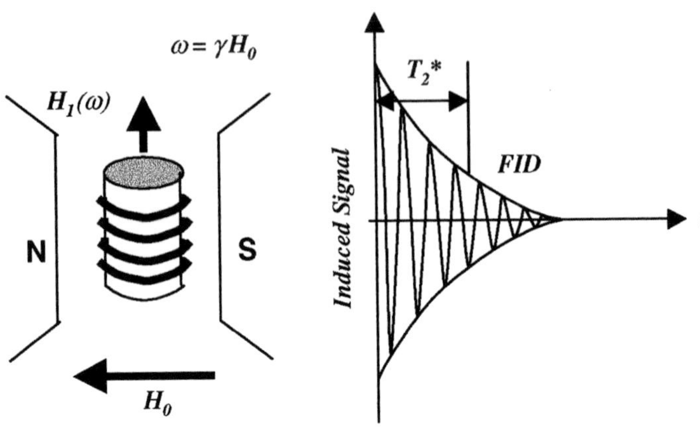

Figure 2.4: In a laboratory set up, the sample is placed inside a coil between the pole pieces of a magnet. The applied field H_0 is from right to left, whereas the rf field H_1 is pointing upward. The induced signal after a 90° pulse is shown as the Free Induction Decay (FID).

2.3 T_1 measurements

In a typical laboratory NMR apparatus, the sample is placed in a test tube/sample holder which is positioned between the pole faces of the magnet (Figure 2.4). An induction coil is looped around the sample holder such that its magnetic field direction is at right angles to the field of the magnet. The coil is energized by pulses of current at the Larmor frequency of the nucleus. The duration of these pulses determines the angle which the spins are turned away from the direction of the magnet. After the pulse is turned off, the same coil or an additional coil is used to measure the decay or relaxation of the signal. The signal in the coil is induced by the precession of the spins about the static field in the transverse, i.e. xy plane. The signal that we measure is called the Free Induction Decay (FID). The FID is largely influenced by T_1, T_2, magnetic field uniformity, and the presence of paramagnetic nuclei. Generally speaking, a long FID, when Fourier transformed to the frequency domain, gives a sharp peak at the Larmor frequency, whereas a short FID

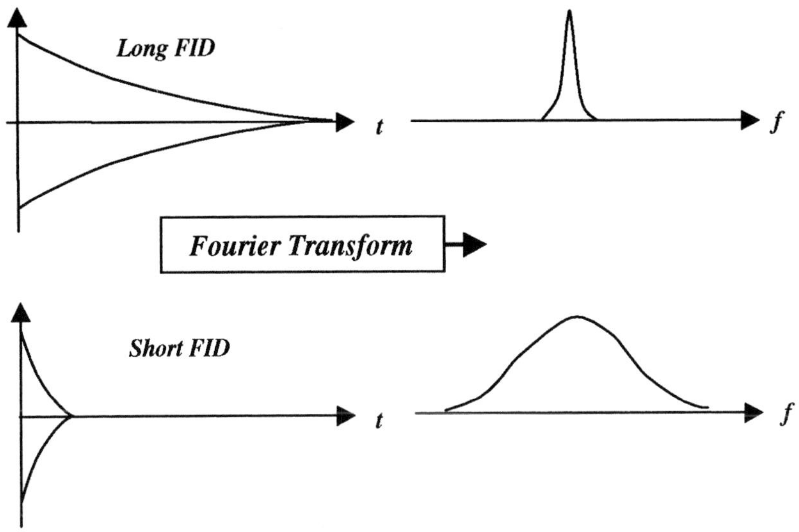

Figure 2.5: A long FID, when Fourier transformed to frequency domain, exhibits a sharp peak at the Larmor frequency, whereas a short FID shows a broad peak, indicating a heterogeneous field distribution over the sample.

exhibits a broad peak, which indicates a heterogeneous field distribution over the volume of the sample, as shown in Figure 2.5. Due to the very heterogeneous field, rock samples typically have a very broad peak in the frequency domain.

Because of the complexity of FID interpretation, multiple-pulse sequences have evolved which separate out these effects. For example, a pulse sequence called "inversion recovery" is used to determine T_1 and a spin-echo sequence (CPMG) is used to determine T_2.

2.3.1 Inversion recovery

The *inversion recovery* pulse sequence has the advantage of being able to measure T_1 and almost eliminate any T_2 effects. The sequence consists of a 180° spin inversion pulse followed by a variable recovery time and then a 90° "read" pulse. The 180° pulse inverts the magnetization vector from the positive z axis (Figure 2.6) to the negative z axis. Because the magnetization vector is entirely longitudinal, there is no transverse component

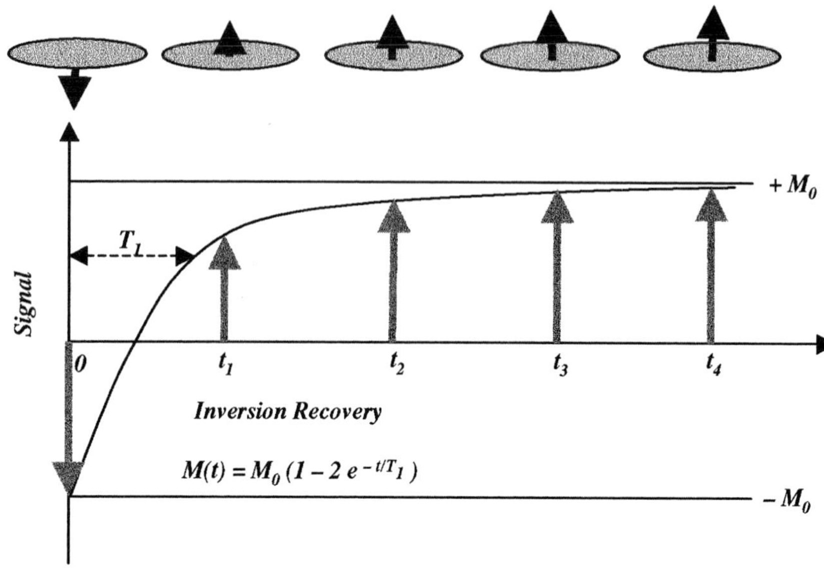

Figure 2.6: A typical sequence of inversion recovery for obtaining T_1.

and, consequently, no measurable signal. Immediately after the stimulus, the spins start to flip back to the positive z direction at the rate of $1/T_1$, which is described by the following equation:

$$M(t) = M_0(1 - 2e^{-t/T_1}). \tag{2.12}$$

How much the vector grows in the $+z$ direction during the variable recovery period also cannot be directly measured because there is no transverse component.

The purpose of the subsequent 90° pulse, then, is to tilt the magnetization into the plane of the receiving coil where it can be measured. A free induction decay (FID) follows the 90° pulse. The strength of the FID is related to the growth of the magnetization along the positive z axis that occurred since the initial 180° pulse.

Figure 2.6 shows a sequence of such experiments in which the recovery times are 0, t_1, t_2, t_3, and t_4, and the corresponding FID is plotted against the variable recovery time. T_1 can then be computed from a fit using Eq. (2.12). A quick estimate of T_1 can be obtained by noting the time t where $M(t)$ crosses the zero value and using Eq. (2.12), i.e., $T_1 = t/(\ln 2)$.

2.3.2 Saturation recovery

There is another pulse sequence, called *saturation recovery*, which uses a 90° pulse followed by a 90° "read" pulse. The magnetization of the saturation recovery sequence is described by the following equation

$$M(t) = M_0(1 - e^{-t/T_1}). \tag{2.13}$$

The popularity of the inversion recovery method is due to the fact that the dynamic range for magnetization is $2M_0$ as opposed to M_0 for the saturation recovery sequence. However, the saturation recovery has the advantage of allowing an arbitrarily short delay between two pulse pairs. In fact, an infinite train of 90° pulses with separation τ can be used. When reaching a steady state, the signal after each pulse is given by

$$M(\tau) = M_0(1 - e^{-\tau/T_1}). \tag{2.14}$$

This is an efficient method for measuring T_1.

2.4 T_2 measurements

2.4.1 Hahn's spin echo

As shown in Figure 2.7, consider a 90° pulse applied to a rock sample in the transverse plane along the x axis. The magnetization vector, which is originally aligned with the upward positive z axis, is deflected 90° into the xy plane, and lies along the y axis. Initially, the transverse vector appears as a single vector because the nuclei are in phase. As time passes after the pulse is turned off, the nuclei "see" variations in the static field and precess at different rates. We observe the transverse component dispersing and fanning out in the xy plane, some nuclei with higher Larmor frequency are moving ahead while those with lower Larmor frequency lag behind.

If at a period of time τ after we applied the 90° pulse, we apply a 180° pulse, all these spins are flipped over to the other side of the xy plane, and continue precessing about the same z axis. Now, because of the "flipping", the phases of spins with higher Larmor frequency now lag, whereas those with lower Larmor frequency now lead. Hence, after exactly a time period of τ, all the spins become in phase again. As a result, we observe a rise in the measured transverse magnetization signal. This phenomenon is called *spin*

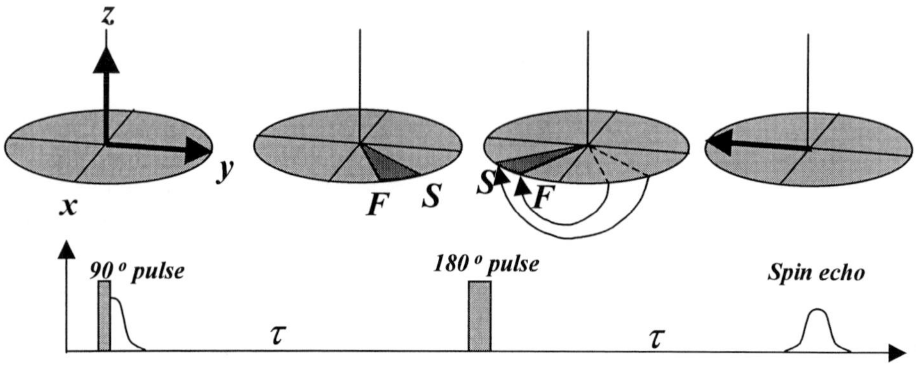

Figure 2.7: A schematic drawing showing the formation of a spin echo.

echo. It was first discovered by Hahn in 1950 [96], and is frequently referred to as *Hahn's echo.*

2.4.2 CPMG measurements

The spin echo produced by the 180° pulse is an echo of the free induction decay produced by the preparative 90° pulse. The echo can be considered a "resurrection" and doubling of the apparently vanished 90° free induction decay. The echo's shape is determined by the shape of the free induction decay curve. After reaching maximum coherence (the top of the echo curve), the nuclei again go out of phase at approximately the same rate as after the initial 90° pulse, the falling half of the echo curve very closely reproduces the original free induction decay curve.

The magnetic field inhomogeneity has two components: one is time independent, and the other time dependent. For the time independent part of the field, non-uniformities in a sample cause T_2^* dephasing to occur at a constant rate. This time independent field heterogeneity usually comes from the applied field **H** and the local field due to susceptibility differences. While the dephasing effects of T_2^* can be reversed by a 180° pulse, those due to time dependent fluctuating fields cannot. The latter are due to neighboring nuclei which are vibrating and rotating and are characterized by the time constant T_2. The echo is thus weaker than the initial amplitude by a factor of $e^{-2\tau/T_2}$.

A train of echoes may be obtained by repeating 180° pulses at appropriate intervals (every 2τ), to produce several echoes after the initial 90° pulse. Each new 180° pulse has the effect of reversing the dephasing that occurred since the previous pulse. In multiple-echo sequences, the height of successive echos decreases exponentially with a time constant T_2. T_2 can therefore be calculated from the decay in the height of the successive echoes, which form a spin echo train described by $e^{-2n\tau/T_2}$. Elaborate pulse sequences have been designed to both remove the T_2^* effects and instrument errors. The most commonly used method is called the CPMG pulse sequence after the successive pairs of authors: Carr and Purcell [54], and Meiboom and Gill [161]. Figure 2.8 shows a CPMG pulse sequence. Theoretically speaking, if there is no diffusion of spins in the system, successive experiments of Hahn's echo taken at 2τ, 4τ, 6τ,..., would be identical to a CPMG spin echo train. Because in reality, spins always undergo diffusive motion, further degradation is caused by using a large τ spacing in the presence of field gradients. Thus, it is preferable to use CPMG with a small τ to reduce the diffusion attenuation of the spin echo. This is discussed in the following section.

In a typical laboratory NMR measurement, a series of CPMG spin echo trains are taken. Signals are stacked and averaged to improve the signal to noise ratio. Between successive CPMG excitations, a time period is inserted to ensure all spins are fully recovered and ready for the next excitation. This time period is referred to as "Repeat Delay", "Wait Time", or "Repeat Time" (and abbreviated as RD, WT, TW, or TR in the literature. In the rest of the book, all these abbreviations are used interchangeably, and they all refer to the same physical quantity). A schematic drawing of such a timing diagram is also shown in Figure 2.8.

2.5 Diffusion in non-uniform fields

2.5.1 Enhanced T_2 relaxation rate due to diffusion

In the above discussion, we have not considered the fact that nuclear spins in a liquid can move away from their initial position by diffusive motion of the atoms. This introduces an additional decay of the spin echo signal if there is

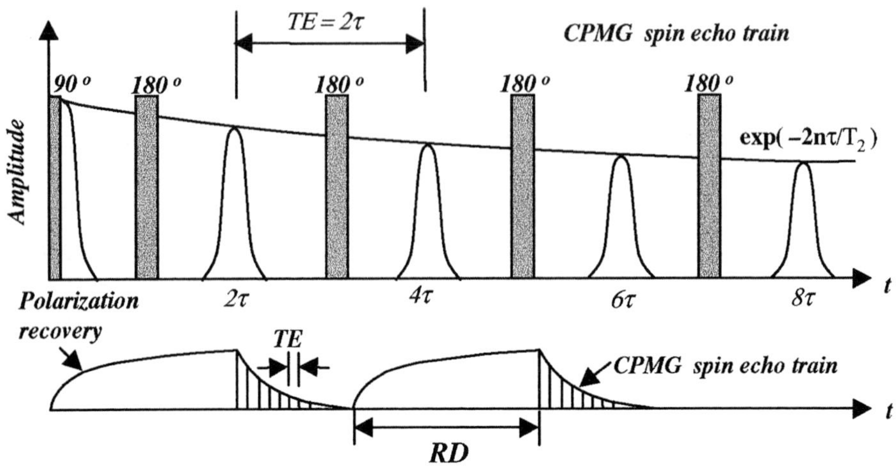

Figure 2.8: After the first spin echo is formed at 2τ, successive 180° pulses are applied at every 2τ with spin echo formed between the pulses. The time between echoes, 2τ, is customarily referred to as TE.

a magnetic field gradient. The decay is characterized by the function

$$\exp\left(-\frac{2\tau}{T_2} - (\gamma \nabla H)^2 \frac{2}{3} D\tau^3\right). \tag{2.15}$$

Here $1/T_2$ is the decay rate of transverse magnetization in a uniform fluid, τ is the time interval used in the dephasing and rephasing periods of the spin echo experiments, D is the self diffusion coefficient of the atom possessing the nuclear spin in the fluid, γ is the nuclear gyromagnetic ratio, and ∇H is the field gradient which is assumed to be uniform.

In a multiple-echo sequence such as CPMG, the time between echoes is 2τ. When considering the nth echo in the echo train, Eq. (2.15) can be written as

$$\exp\left(-\frac{2n\tau}{T_2} - (\gamma \nabla H)^2 \frac{2}{3} Dn\tau^3\right), \tag{2.16}$$

where $2n\tau$ is simply the decay time t. Hence the total relaxation rate is given by

$$\text{Rate} = \frac{1}{T_2} + (\gamma \nabla H)^2 D \frac{\tau^2}{3}. \tag{2.17}$$

Thus, if we choose a very small τ, we can make the diffusion term as small as possible. The purpose of the Carr-Purcell-Meiboom-Gill (CPMG) pulse sequence is to remove the effects of gradients in the static magnetic field on the T_2 relaxation rate by using a small τ.

Note that this enhanced relaxation rate due to diffusion in a field gradient only affects T_2, not T_1[1]. This is because diffusion causes phase incoherency for spins precessing in the xy plane, directly impacting T_2. The T_1 relaxation, on the other hand, has to do with the rate of energy exchange between the magnetic energy of the nuclei and the thermal energy of the vibrating molecules in the system.

In NMR logging, CPMG is the primary pulsing sequence used for logging, and T_2 is the primary quantity measured for extracting relevant information. Because the magnetic fields generated by the NMR logging tools usually have large field gradients, Eqs. (2.15-2.17) are often used extensively and freely in NMR log interpretation to differentiate oil and water, primarily using the fact that there are large differences in their diffusion constants. However, one should keep in mind that Eqs. (2.15-2.17) are only strictly valid for fluids of infinite extent and in uniform gradients, and are approximately valid for fluids in porous media under limited conditions. This is further discussed in Chapter 7.

To derive Eq. (2.15)[2], we recall the Bloch equations for the decays of M_x and M_y in the xy-plane of the rotating frame:

$$\frac{dM_x}{dt} = -\frac{M_x}{T_2}, \qquad \frac{dM_y}{dt} = -\frac{M_y}{T_2}. \tag{2.18}$$

For the consideration of the diffusion effect, an additional diffusion term needs to be included:

$$\frac{dM_x}{dt} = -\frac{M_x}{T_2} + D\nabla^2 M_x, \qquad \frac{dM_y}{dt} = -\frac{M_y}{T_2} + D\nabla^2 M_y \tag{2.19}$$

where D is the diffusion constant, and $D\nabla^2 M = \nabla \cdot D\nabla M$ is the net diffusion influx which contributes to the time dependent quantity, dM/dt. Introducing

[1] Strictly speaking, the statement is true in principle, and is true in practice only when $\nabla H \cdot \Delta\mathbf{r} \ll H_1$, where $\Delta\mathbf{r}$ is the distance of spin diffusion during the measurements, and H_1 is the rf field strength. If the condition is not satisfied, spins may diffuse out of resonance condition, $\Delta\omega \approx \gamma H_1$, then the T_1 measurement will be affected.

[2] The derivation here follows mainly that of Slichter [201].

$M^+ \equiv M_x + iM_y$, we can rewrite the previous equations in a concise form:

$$\frac{dM^+}{dt} = -\frac{M^+}{T_2} + D\nabla^2 M^+. \tag{2.20}$$

Let us assume we have a field distribution as follows

$$H(\mathbf{r}) = H_0 + \mathbf{r} \cdot \nabla H \tag{2.21}$$

where $H(\mathbf{r})$ is the time independent field along the z axis, H_0 is the averaged field, and ∇H is the field gradient which we assume is constant.

Recall that the magnetization M, once tilted into xy plane, starts to precess about the z axis. Since the magnetization is nothing more than the sum of all spins, μ, in the system, the equation describing the precession of the magnetization in the laboratory (stationary) frame is similar to Eq. (2.3), i.e.,

$$\frac{dM}{dt} = M \times (\gamma \mathbf{H}). \tag{2.22}$$

Eq. (2.20) is for a rotating frame of reference. This frame rotates about the z axis at the Larmor angular frequency, $\omega_0 = \gamma H_0$. Thus, if the field were perfectly homogeneous with a value H_0, all spins would be precessing at exactly the same Larmor angular frequency, ω_0, and the magnetization would be stationary in the rotating frame. There would not be any time dependent contribution to Eq. (2.20) from the precessing of the magnetization in the laboratory frame.

However, the field is *not* homogeneous, and there are regions with higher and lower Larmor frequencies. These spins with differences in Larmor frequencies

$$\gamma \Delta H(\mathbf{r}) = \gamma (H(\mathbf{r}) - H_0) = \gamma \mathbf{r} \cdot \nabla H \tag{2.23}$$

contribute an extra term

$$M^+ \times \gamma \Delta H = -i\gamma (\mathbf{r} \cdot \nabla H) M^+ \tag{2.24}$$

to Eq. (2.20) which describes the net precession of leading or lagging spins in the rotating frame. Hence Eq. (2.20) now becomes:

$$\frac{dM^+}{dt} = -i\gamma (\mathbf{r} \cdot \nabla H) M^+ - \frac{M^+}{T_2} + D\nabla^2 M^+ \tag{2.25}$$

which is known as Torrey's equation [215].

If there were no diffusion, i.e., D=0, Eq. (2.25) would have the following solution

$$M^{+}(\mathbf{r},t) = M_0 e^{-t/T_2} e^{-i\gamma(\mathbf{r}\cdot\nabla H)t}, \tag{2.26}$$

where M_0 is the magnetization of the system at t=0, right after it is tilted to the xy plane following the 90° pulse. If the field were perfectly homogeneous, the dephasing term, $e^{-i\gamma(\mathbf{r}\cdot\nabla H)t}$, would not be there, and the signal would decay simply as e^{-t/T_2}. With the field inhomogeneity, the dephasing term, $e^{-i\gamma(\mathbf{r}\cdot\nabla H)t}$, is present, and causes the signal to decay much faster, resulting in a much shorter decay time constant T_2^*. However, if after time τ, a 180° pulse is applied, reversing the sign of the dephasing angle, i.e., becoming $e^{i\gamma(\mathbf{r}\cdot\nabla H)\tau}$, the spins would begin to refocus, resulting in the resurrection of a spin echo at exactly a time τ after the 180° pulse. Thus, a CPMG-type spin echo train would again decay as e^{-t/T_2}.

Under the influence of diffusion in an inhomogeneous field, further signal degradation occurs. Let us consider the magnetization density at $\mathbf{r} - \Delta\mathbf{r}$, $\mathbf{r}$, and $\mathbf{r}+\Delta\mathbf{r}$, where $\Delta\mathbf{r}$ is a small distance. At time t=0, the spins at these three locations have the same phase. As the time progresses, the magnetization at $\mathbf{r}+\Delta\mathbf{r}$ develops a lead angle of $\gamma(\Delta\mathbf{r}\cdot\nabla H)t$ over the magnetization at $\mathbf{r}$. Similarly, the magnetization at $\mathbf{r}-\Delta\mathbf{r}$ develops the same amount of lag angle behind the magnetization at $\mathbf{r}$. During this time, the spin diffusion occurs. Since the distance between $\mathbf{r}+\Delta\mathbf{r}$ and $\mathbf{r}$, is the same as that between $\mathbf{r}$ and $\mathbf{r} - \Delta\mathbf{r}$, equal amounts of spins with lead and lag angles would diffuse to $\mathbf{r}$. Thus, the mean phase at $\mathbf{r}$ will remain unchanged. However, the magnitude of the signal at $\mathbf{r}$ will decrease, because spins with coherent phase are moving away from $\mathbf{r}$, whereas those moving into $\mathbf{r}$ are not in phase. This argument can be applied to any point $\mathbf{r}$ in space, as long as it is not near the boundary. We thus expect that the diffusion-induced decay will not affect the phase angle, and is independent of $\mathbf{r}$ but dependent on time t. Therefore, we try a solution of the following form:

$$M^{+}(\mathbf{r},t) = M_0 e^{-t/T_2} e^{-i\gamma(\mathbf{r}\cdot\nabla H)t} A(t) \tag{2.27}$$

When this is substituted in Eq. (2.25), we get

$$\frac{1}{A}\frac{dA}{dt} = -D(\gamma\nabla H)^2 t^2 \tag{2.28}$$

with the following solution

$$A(t) = A(0)\exp\left[-D(\gamma\nabla H)^2\frac{t^3}{3}\right]. \tag{2.29}$$

Lumping $A(0)$ into M_0, we now have

$$M^+(\mathbf{r},t) = M_0 e^{-t/T_2} e^{-i\gamma(\mathbf{r}\cdot\nabla H)t} \exp\left[-D(\gamma\nabla H)^2\frac{t^3}{3}\right] \tag{2.30}$$

for the temporal behavior of the magnetization following the initial 90° pulse, when the diffusion effect is included.

We now consider the formation of a spin echo using Eq. (2.30). At time τ after the application of a $\pi/2$ pulse and just prior to the application of the π pulse, the magnetization has the following form:

$$M^+(\mathbf{r},\tau^-) = M_0 e^{-\tau/T_2} e^{-i\gamma(\mathbf{r}\cdot\nabla H)\tau} \exp\left[-D(\gamma\nabla H)^2\frac{\tau^3}{3}\right]. \tag{2.31}$$

The application of a π pulse has the effect[3] of replacing $-\gamma(\mathbf{r}\cdot\nabla H)\tau$ with $\gamma(\mathbf{r}\cdot\nabla H)\tau$. Thus, immediately after the π pulse, the magnetization has the following form:

$$M^+(\mathbf{r},\tau^+) = M_0 e^{-\tau/T_2} e^{i\gamma(\mathbf{r}\cdot\nabla H)\tau} \exp\left[-D(\gamma\nabla H)^2\frac{\tau^3}{3}\right]. \tag{2.32}$$

From this time, the further development of $M^+(\mathbf{r},t)$ will follow Eq. (2.30), i.e., continuing the precession about z-axis:

$$M^+(\mathbf{r},\tau+t) = M^+(\mathbf{r},\tau^+) e^{-t/T_2} e^{-i\gamma(\mathbf{r}\cdot\nabla H)t} \exp\left[-D(\gamma\nabla H)^2\frac{t^3}{3}\right]. \tag{2.33}$$

At exactly τ after the application of the π pulse, i.e., $t=\tau$, we have

$$M^+(\mathbf{r},2\tau) = M^+(\mathbf{r},\tau^+) e^{-\tau/T_2} e^{-i\gamma(\mathbf{r}\cdot\nabla H)\tau} \exp\left[-D(\gamma\nabla H)^2\frac{\tau^3}{3}\right]. \tag{2.34}$$

Substituting $M^+(\mathbf{r},\tau^+)$ from (2.32) into (2.34), we note that the complex phase factors, $e^{-i\gamma(\mathbf{r}\cdot\nabla H)\tau}$ and $e^{i\gamma(\mathbf{r}\cdot\nabla H)\tau}$, cancel each other, resulting in the formation of a spin echo at 2τ

$$M^+(\mathbf{r},2\tau) = M_0 e^{-2\tau/T_2} \exp\left[-D(\gamma\nabla H)^2\frac{2\tau^3}{3}\right]. \tag{2.35}$$

[3] The derivation here is slightly different from that of Slichter's [201], or Torrey's [215] derivation. In their derivation, the initial $\pi/2$ pulse and the subsequent successive π pulses are all applied along x axis. Thus, the application of a π pulse has the effect of replacing $-\gamma(\mathbf{r}\cdot\nabla H)\tau$ with $\pi+\gamma(\mathbf{r}\cdot\nabla H)\tau$, resulting in a $(-)^n$ factor in the CPMG echo trains. Here, we apply the subsequent successive π pulses along y axis. Thus, the application of a π pulse has the effect of replacing $-\gamma(\mathbf{r}\cdot\nabla H)\tau$ with $\gamma(\mathbf{r}\cdot\nabla H)\tau$.

This process of π pulse application can be repeated, and each diffusion loss takes place independent of the previous cycle, as long as it is not near the boundary of the system. Thus, after n cycles, we get

$$M^{+}(\mathbf{r}, 2n\tau) = M_0 e^{-2n\tau/T_2} \exp\left[-D(\gamma\nabla H)^2 \frac{2n\tau^3}{3}\right], \tag{2.36}$$

which is the CPMG spin echo train.

2.5.2 Pulsed field gradient spin echo

Based on Eq. (2.15), one should be able to measure the magnetization decay as a function of varying τ or varying field gradient ∇H to obtain the self diffusion coefficient D of the fluid. However, there are experimental limitations in the ordinary spin-echo experiment, which requires the magnetic field gradient to be present all the times. Stejskal and Tanner [205] came up with a pulsed field gradient spin echo technique which provides a better control of the experimental parameters and a more clearly defined diffusion time. Their method has since been widely used for diffusion coefficient measurements.

Instead of using a constantly "on" magnetic field gradient, Stejskal and Tanner suggested that a pulsed field gradient, significantly larger than the background field gradient, be turned on for a short duration to perform the dephasing and rephasing of the spins in the spin echo experiment. Their pulsing sequence is shown in Figure 2.9. The variation of the field gradient as a function of time is described as follows:

$$\begin{aligned} \mathbf{g}_0 &\quad \text{when} \quad 0 < t < t_1, \\ \mathbf{g}_0 + \mathbf{g} &\quad \text{when} \quad t_1 < t < t_1 + \delta < \tau, \\ \mathbf{g}_0 &\quad \text{when} \quad t_1 + \delta < t < t_1 + \Delta > \tau, \\ \mathbf{g}_0 + \mathbf{g} &\quad \text{when} \quad t_1 + \Delta < t < t_1 + \Delta + \delta < 2\tau, \\ \mathbf{g}_0 &\quad \text{when} \quad t_1 + \Delta + \delta < t, \end{aligned} \tag{2.37}$$

where $\mathbf{g}_0$ is the background field gradient which is present at all times, $\mathbf{g}$ is the pulsed field gradient which is turned on for a short duration of time δ, once between the 90° initial pulse and the 180° pulse and once again between the 180° pulse and the echo, and Δ is the diffusion time between the successive gradient pulses. Since the pulsed field gradient $\mathbf{g}$ is significantly larger than the background field gradient $\mathbf{g}_0$ and the diffusion time Δ is well-defined, the measurement of D is more accurate.

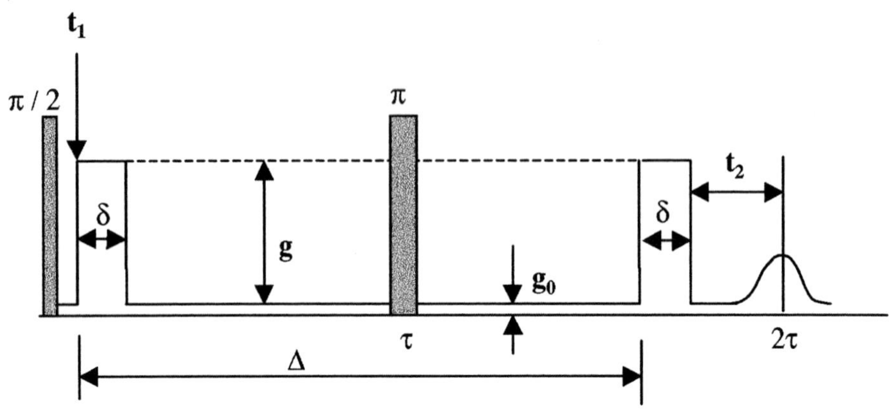

Figure 2.9: The PFGSE pulsing diagram of the Stejskal and Tanner sequence [205], where $\mathbf{g_0}$ is a constant background field gradient and $\mathbf{g}$ is the applied field gradient.

Following the same procedure as shown in the previous section, it can be shown, after many algebraic manipulations, that the spin echo amplitude is given by:

$$\ln[M(2\tau)/M_0] = -\gamma^2 D\{\tfrac{2}{3}\tau^3 g_0^2 + \delta^2(\Delta - \tfrac{1}{3}\delta)g^2 - \delta[(t_1^2 + t_2^2) + \delta(t_1 + t_2) + \tfrac{2}{3}\delta^2 - 2\tau^2]\mathbf{g}\cdot\mathbf{g_0}\} \qquad (2.38)$$

where $t_2 = 2\tau - (t_1 + \Delta + \delta)$ is the time between the end of the second field gradient pulse and the peak of the echo.

When $\mathbf{g} = 0$, Eq. (2.38) reduces to Eq. (2.15).

Tanner [210] suggested a stimulated echo[4] pulse sequence for measuring the diffusion coefficient where the T_2 of the system is much less than the T_1. The scheme employs a sequence of three 90° pulses as shown in Figure 2.10, where both τ_1 and τ_2 are measured after the first 90° pulse, and $t_2 \equiv \tau_1 +$

[4]A stimulated echo occurs at $\tau_1 + \tau_2$ after three 90° pulses. It was described in Hahn's paper on spin echoes [96]. There are additional echoes at $2\tau_2 - 2\tau_1$, $2\tau_2 - \tau_1$, and $2\tau_2$ when $2\tau_1 < \tau_2 < T_2$. For $\tau_1 < \tau_2 < 2\tau_1$, the signal at $2\tau_2 - 2\tau_1$ is missing, but the others remain.

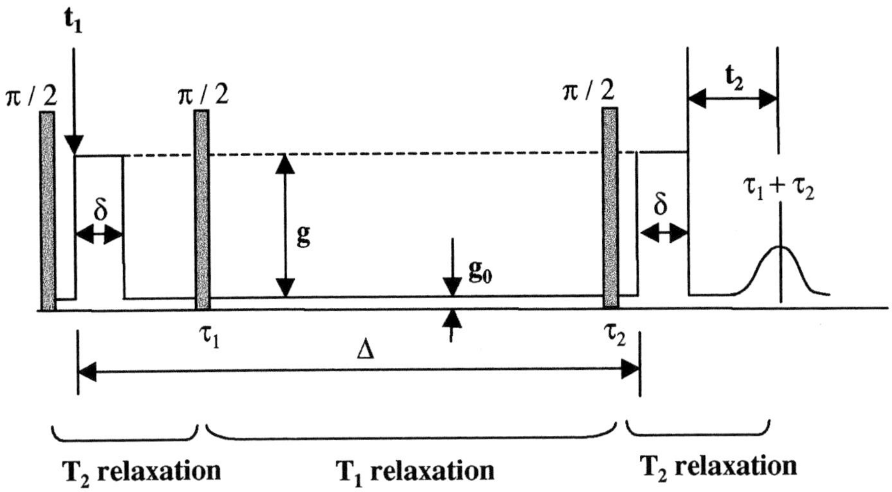

Figure 2.10: The PFGSE pulsing diagram of the Tanner sequence [210] for the stimulated echo, where three $\pi/2$ pulses are used, and $\mathbf{g_0}$ is a constant background field gradient and $\mathbf{g}$ is the applied field gradient.

$\tau_2 - (t_1 + \Delta + \delta)$. The relaxation attenuation of the signal during the interval between the second and the third 90° pulses has a T_1 dependence. Thus the signal strength is much stronger than the regular 90°-180° experiment where the relaxation attenuation has a T_2 dependence.

The variation of the field gradient as a function of time is described as follows:

$$
\begin{array}{rll}
\mathbf{g_0} & \text{when} & 0 < t < t_1, \\
\mathbf{g_0} + \mathbf{g} & \text{when} & t_1 < t < t_1 + \delta < \tau_1, \\
\mathbf{g_0} & \text{when} & t_1 + \delta < t < t_1 + \Delta > \tau_2, \\
\mathbf{g_0} + \mathbf{g} & \text{when} & t_1 + \Delta < t < t_1 + \Delta + \delta < \tau_1 + \tau_2, \\
\mathbf{g_0} & \text{when} & t_1 + \Delta + \delta \equiv \tau_1 + \tau_2 - t_2 < t,
\end{array}
\tag{2.39}
$$

and the echo amplitude is given by:

$$
\begin{aligned}
\ln[M(2\tau)/M_0] = -\gamma^2 D\{\delta^2(\Delta - \tfrac{1}{3}\delta)g^2 + \tau_1^2(\tau_2 - \tfrac{1}{3}\tau_1)g_0^2 \\
- \delta[(t_1^2 + t_2^2) + \delta(t_1 + t_2) + \tfrac{2}{3}\delta^2 - 2\tau_1\tau_2]\mathbf{g} \cdot \mathbf{g_0}\}.
\end{aligned}
\tag{2.40}
$$

If we set $\tau_1 = \tau_2 = \tau$, i.e., the time between τ_1 and τ_2 becomes zero and the two $\pi/2$ pulses collapse into a single π pulse just as that shown in Figure 2.9, then Eq. (2.40) reduces to Eq. (2.38).

2.6 Practical measurement problems

In the discussion so far in this Chapter on the measurements of T_1 and T_2, we have assumed an ideal situation where a perfect 90° or 180° can be achieved, and the signal is optimized. This is possible in a laboratory environment where the sample is placed in between the pole pieces of a magnet and is inside a rf coil, as shown in Figure 2.4, where homogeneous applied field and rf field are possible.

In practical NMR logging measurements, however, the probed formation is outside the tool housing and far away from the magnet, as shown in Figures 1.3-1.5. The magnetic field over the probed zone is far from homogeneous. The rf field which tips the spin magnetization is also not uniform. As a result, the effective $\mathbf{H_1}$ is not exactly perpendicular to the applied field $\mathbf{H_0}$ along the z axis, and the optimal rf pulses are not exactly 90° or 180°. In fact, they can be far from those values.

There are many practical problems associated with NMR measurements. We shall only focus on those problems encountered in NMR logging measurements. The magnetic fields over the probed formation usually have large gradients. The effects of the field inhomogeneity on the CPMG echo trains are quite complicated. We shall simply illustrate the conceptual approach for solving these problems without going into the details. Interested readers are referred to the literature on this subject [94, 227, 68, 13, 14, 111].

2.6.1 Inhomogeneous rf field

When the rf field, i.e., $\mathbf{H_1}$, is not homogeneous, not all the spins turn the same 90° or 180° after the application of the rf pulses. As shown in Figure 2.11, we illustrate the case when the "π" pulse is less than π. The $\mathbf{H_1}$ field is along the x axis. The initial $\pi/2$ pulse tips the magnetization from the z axis into the y axis. After a time τ, the phase of the precessing spins spread out due to inhomogeneity in the applied field $\mathbf{H_0}$. The fast and slow precessing spins are indicated by F and S, respectively. When a slightly less than π (i.e., $\pi - \delta$) pulse is applied, the spins are flipped over to a circle slightly above

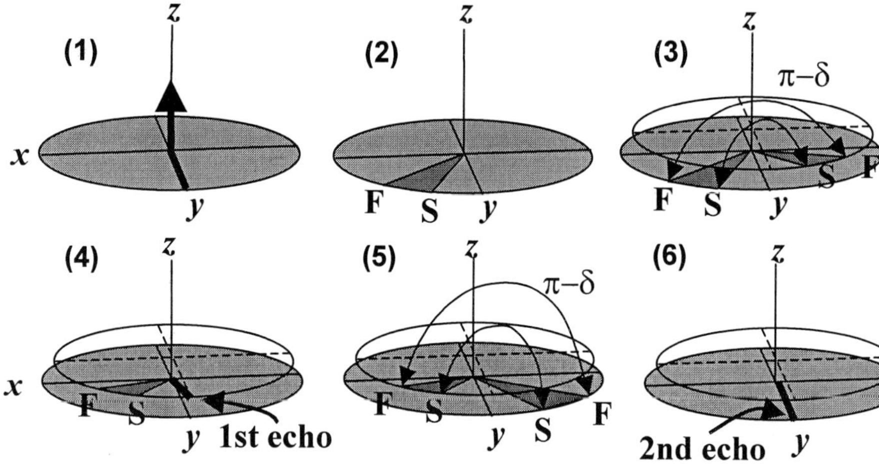

Figure 2.11: Schematic diagram showing the first and second echoes after the application of a $\pi - \delta$ pulse which is less than π. The $\mathbf{H_1}$ field is along x axis. (1) The initial $\pi/2$ pulse tips the magnetization from z into y axis. (2) The precessing spins spread out due to the inhomogeneity of H_0. (3) A $\pi - \delta$ pulse flips the precessing spins to a circle above the xy plane. (4) After the first echo is formed, the precessing spins spread out again. (5) The second $\pi - \delta$ pulse flips the precessing spins back to the xy plane. (6) The second echo is formed with a normal amplitude.

the xy plane. After a time τ, the first spin echo is formed, the amplitude of which is smaller than if the "π" pulse were exact. When the second $\pi - \delta$ pulse is applied after another time τ, the precessing spins are brought back to the xy plane. Thus the second spin echo has the normal amplitude when it is formed.

The same situation occurs when the pulse is larger than π. Now the precessing spins will be flipped over to a circle below the xy plane after the first $\pi + \delta$ pulse application, and brought back to the xy plane after the second $\pi + \delta$ pulse application. Thus, the general symptom of an imperfect π pulse for a CPMG echo train is to have an oscillatory appearance, with the odd echoes smaller than what they should be. One solution would be to use only the even echoes for data analysis.

2.6.2 Inhomogeneous H_0 field

When the applied field $\mathbf{H_0}$ in not homogeneous, there will be a distribution of Larmor frequencies, i.e., $\omega_0(\mathbf{r}) = \gamma H_0(\mathbf{r})$, which is a function of the spatial coordinates. If the rf field frequency ω_{rf} is tuned to the average Larmor frequency, then there will be an offset $\Delta\omega_0(\mathbf{r})$ given as follows:

$$\Delta\omega_0(\mathbf{r}) \equiv \gamma H_0(\mathbf{r}) - \omega_{\text{rf}}. \tag{2.41}$$

As shown in Figure 2.12, in the rotating frame with angular velocity ω_{rf}, if all spins had the same Larmor angular frequency ω_{rf} (i.e., $\Delta\omega_0(\mathbf{r}) = 0$ everywhere), the rf field $\mathbf{H_1}$ would be lying exactly along the x axis. Because quite a few spins have $\Delta\omega_0 \neq 0$, they precess about the z axis in the rotating frame with angular velocity of $\Delta\omega_0$. This is equivalent to having an effective field of $\Delta H_0 = \Delta\omega_0/\gamma$ along the z axis. Thus, when the initial $\pi/2$ pulse is applied, the magnetization along the z axis rotates about an effective rf field, denoted by $\mathbf{H_{1eff}}$ in Figure 2.12, which is the sum of ΔH_0 and H_1. The angle of rotation

$$\theta = \frac{\sqrt{H_1^2 + (\Delta H_0)^2}}{H_1}\frac{\pi}{2} \tag{2.42}$$

is slightly larger than $\pi/2$, where the pulse width is determined by $\pi/(2\gamma H_1)$ when $\Delta\omega_0$ is zero. The subsequent applications of "π" pulses about the y axis suffer from the same effect. The trajectory of the tip of the magnetization from the application of the initial "$\pi/2$" pulse through the formation of the first and second echoes is shown in Figure 2.12. The locations of the first and second echoes are denoted by solid dots in the figure. The loci of subsequent echoes occur between these two dots.

2.6.3 Spin dynamics in inhomogeneous fields

The above discussion treated the effects of inhomogeneous $\mathbf{H_1}$ and $\mathbf{H_0}$ fields individually. In fact, they usually happen concurrently in NMR logging measurements. There is a distribution of $\Delta\omega_0$ as well as a distribution of $\omega_1 = \gamma H_1$. More elaborate considerations can be found in [94, 111]. Here, we shall simply summarize the general approach without going into the mathematical details.

As discussed previously, because of the inhomogeneity of the $\mathbf{H_0}$ field, the rf field will now have an effective field, $\mathbf{H_{1eff}}$, with a different orientation and

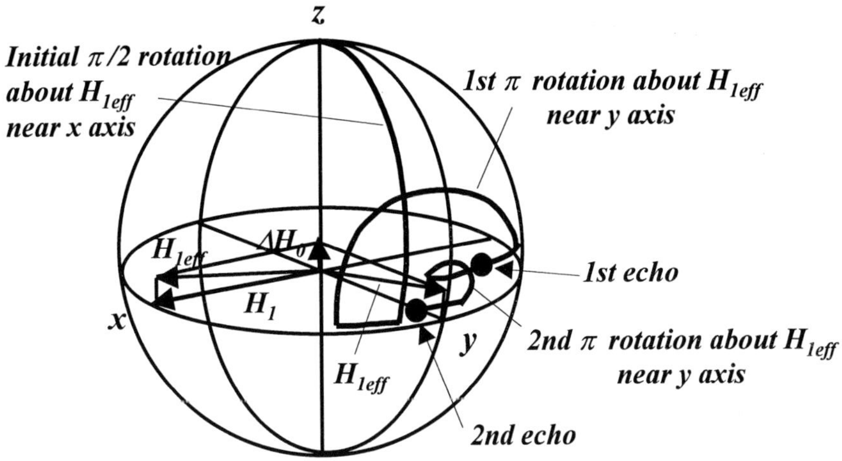

Figure 2.12: Schematic diagram showing the effect of an inhomogeneous $\mathbf{H_0}$ field in a rotating frame. The rf field H_1 combined with the offset ΔH_0 form an effective field $\mathbf{H_{1eff}}$ slightly tilted above the xy plane. The trajectory of the tip of the magnetization from the application of the initial "$\pi/2$" pulse through the formation of the first and second echoes is shown as a heavy line.

magnitude. The effective axis $\hat{n}$ and magnitude $H_{1\text{eff}}$ are given as follows:

$$
\begin{aligned}
H_{1\text{eff}} &= \sqrt{H_1^2 + (\Delta H_0)^2} = \Omega/\gamma & (2.43)\\
\Omega &= \sqrt{\omega_1^2 + \Delta\omega_0^2} & (2.44)\\
\hat{n} &= \frac{\omega_1}{\Omega}(\ \hat{x}\cos\phi + \hat{y}\sin\phi\ ) + \frac{\Delta\omega_0}{\Omega}\hat{z} & (2.45)
\end{aligned}
$$

where $\hat{x}$, $\hat{y}$, and $\hat{z}$ are unit vectors along the x, y, and z axes, respectively. ϕ is the phase angle for the rf field. The rf field is pointing along x axis when $\phi = 0$, and along y axis when $\phi = \pi/2$ for $\Delta\omega_0 = 0$.

Thus, to track the spin dynamics in a rotating frame as that shown in Figure 2.12, the matrix of rotation, $\mathcal{R}_{\pi/2}(\theta, \hat{n}_\alpha)$ for the initial $\pi/2$ pulse can be written as:

$$\mathcal{R}_{\pi/2}(\theta, \hat{n}_\alpha) = \begin{bmatrix} \cos\theta\cos^2\alpha + \sin^2\alpha & -\sin\theta\cos\alpha & (1-\cos\theta)\sin\alpha\cos\alpha \\ \sin\theta\cos\alpha & \cos\theta & -\sin\theta\sin\alpha \\ (1-\cos\theta)\sin\alpha\cos\alpha & \sin\theta\sin\alpha & \cos\theta\sin^2\alpha + \cos^2\alpha \end{bmatrix} \tag{2.46}$$

where $\hat{n}_\alpha$ is the axis of rotation given by Eq. (2.45) when $\phi = 0$, θ is the angle of rotation, and

$$\theta = \frac{\sqrt{H_1^2 + (\Delta H_0)^2}}{H_1}\frac{\pi}{2} \qquad \text{and} \qquad \alpha = \tan^{-1}\frac{H_1}{\Delta H_0}. \tag{2.47}$$

Similarly, the matrix of rotation for the subsequent "π" pulse about the y axis can be denoted by $\mathcal{R}_\pi(\vartheta, \hat{n}_\beta)$, where $\hat{n}_\beta$ would be the effective axis of rotation given by Eq. (2.45) when $\phi = \pi/2$, and ϑ the angle of rotation, given by

$$\vartheta = \frac{\sqrt{H_1^2 + (\Delta H_0)^2}}{H_1}\pi. \tag{2.48}$$

Thus, the overall evolution after the initial $\pi/2$ pulse until the formation of the first spin echo consists of the following operations

$$\mathcal{R}(\Delta\omega_0\tau, \hat{z})\mathcal{R}_\pi(\vartheta, \hat{n}_\beta)\mathcal{R}(\Delta\omega_0\tau, \hat{z}), \tag{2.49}$$

where $\mathcal{R}(\Delta\omega_0\tau, \hat{z})$ represents the free precession of the spin about z axis during the time period τ when the rf field is off. These operations can be regarded as a single operation denoted by $\mathcal{P} \equiv \mathcal{R}(\Delta\omega_0\tau, \hat{z})\mathcal{R}_\pi(\vartheta, \hat{n}_\beta)\mathcal{R}(\Delta\omega_0\tau, \hat{z})$. It is repeated N times for the formation of the N-th echo. The overall operation to reach the N-th echo can be written as

$$\mathbf{m}_N = \mathcal{P}^N\mathcal{R}_{\pi/2}(\theta, \hat{n}_\alpha) \tag{2.50}$$

where $\mathbf{m}_N$ is the magnetization for a specific $\Delta\omega_0(\mathbf{r})$ after the N-th cycle.

With this, we can now integrate Eq. (2.50), which is a function of both $\Delta\omega_0$ and ω_1, over the whole probed zone and compute all echo amplitudes for the inhomogeneous $\mathbf{H_0}$ and $\mathbf{H_1}$ fields. The foregoing spin dynamics calculation assumes $T_1 = T_2 = \infty$. However, finite T_1 and T_2 relaxations as well as the diffusion effect can be included in the computation. Numerical computations for some of the NMR logging tools have been reported [94, 111]. They show that the field gradients typically have the effect of suppressing the amplitude of the first echo, slightly enhancing the amplitude of the second echo, and diminishing influences on subsequent echoes.

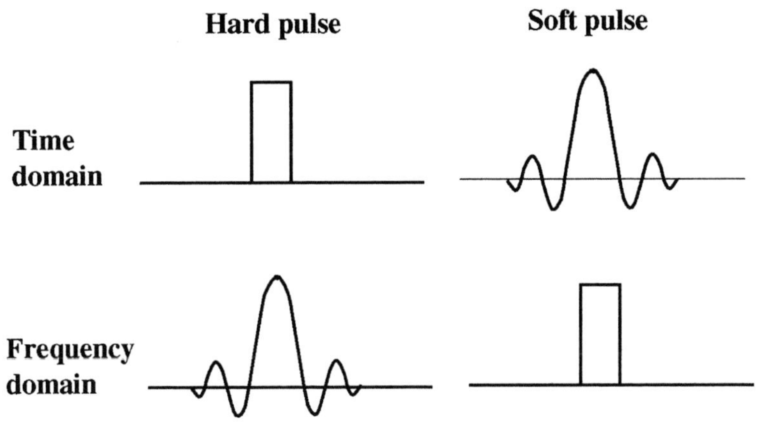

Figure 2.13: Schematic diagram showing the hard and soft pulses in time and frequency domains.

2.6.4 Selection of pulses and flip angles

There are other practical considerations in NMR logging measurements, such as the selection of pulse shapes for multiple frequency excitations, and the selection of flip angles for optimizing the power consumption and signal to noise ratio.

Typically, a hard pulse, i.e., a rectangular shape pulse, is used for the rf excitation. As shown in Figure 2.13, the Fourier transform of the hard pulse in time domain is a sinc function in the frequency domain, whereas the Fourier transform of a soft pulse, i.e., a sinc function in the time domain, has a rectangular shape in the frequency domain. In a logging tool where multiple rf frequency excitations [55] are possible, frequently a soft pulse is used to have a well-defined excited region in the frequency domain so as to avoid interfering with the selected neighboring regions with slightly different excitation frequencies. This frequency hopping increases the efficiency of logging measurements. On the other hand, soft pulses excite spins in a smaller sensitive volume than the hard pulses and thus produce an inherently lower signal.

The selection of flip angle also impacts the power consumption for energizing the rf field as well as the overall signal to noise ratio of the measurements. A large flip angle consumes more rf energy than that of a small flip angle. The inhomogeneity of the $\mathbf{H_0}$ field also has to be taken into account in the overall optimization of signal to noise ratio. Numerical calculations [185, 26] have shown that the optimal flip angles can be significantly less than 90° and 180°.

Chapter 3

NMR properties of fluids

3.1 Introduction

In designing a down-hole NMR logging tool in the 1950s, the researchers initially assumed that the hydrogen nuclei in oil and brine saturated rocks would relax at their bulk rates. The differences in relaxation rate would then be used to identify the oil and brine fractions in the rocks. Because the amplitude of NMR relaxation measurements was proportional to the amount of hydrogen sensed, they presumed also that the tool would be able to determine formation porosity.

Subsequent studies showed that the NMR relaxation of brine in rocks is heavily dominated by surface relaxation. Unrelated studies also showed that the T_2 signal was subject to an additional relaxation enhancement due to diffusion of the nuclei in a magnetic field gradient. Thus, to interpret relaxation measurements in fluid saturated rocks, we need to sort out the contributions from bulk and surface relaxation as well as the diffusion effect. We shall start with NMR relaxation measurements on the bulk fluids typically found in hydrocarbon reservoirs in this chapter, and then cover surface relaxation and diffusion effects of these fluids in porous media in Chapter 4.

Saturating Fluids

In NMR studies of porous media for petrophysics, the saturating fluids of interest are water (of variable salinity), gas, and oil. The water salinity can range from nearly pure to salt saturated (most commonly the salt is NaCl). The gas we discuss is the free gas, mainly methane, i.e., in gaseous form, rather than dissolved in oils. The oil can be free of gas (dead oil) or gas saturated (live oil). Live oils exist at elevated temperatures and pressures where the gas dissolved in the oil is below its bubble point. Synthetic oils used in drilling muds can invade the reservoir either displacing or mixing with the reservoir fluids, and are thus also of interest to the oil industry.

Hydrogen Index

Using NMR logging tools to determine the porosity of a hydrocarbon reservoir is based on calibrations of the tools with respect to known volumes of water. Such calibrations require that the NMR signal per unit volume of each saturating fluid be known. Thus it is convenient to introduce the term hydrogen index (HI) to quantify the atomic hydrogen densities of oil or gas relative to that of an equal volume of water. The hydrogen index is defined as the amount of hydrogen per unit volume of a fluid divided by the amount of hydrogen per unit volume of pure water in standard conditions, i.e.,

$$\mathrm{HI} = \frac{\text{Amount of Hydrogen in Sample}}{\text{Amount of Hydrogen in an equal volume of Pure Water}}. \tag{3.1}$$

Thus hydrogen index should be a quantity independent of measurement methods. Sometimes, when there are limitations in measurement methods, *apparent* rather than *true* hydrogen index is measured.

Oil field brines exhibit a wide variation in salinity. At low salinity, the HI of brine is approximately 1, but at high salinity, the HI can be reduced by as much as 10% (see Figure 3.1), as a small but significant fraction of the brine volume is taken up by the dissolved salt.

The hydrogen indices of most "dead oils" are very close to 1, but the hydrogen index of gas is strongly dependent on its pressure. When the density (ρ), the number of hydrogen atoms in the chemical formula (n_H), and the molecular weight (M_W) of the oil or gas are known, the following equation can be used to calculate HI at room conditions

$$\mathrm{HI} = \frac{9\rho n_H}{M_W}, \tag{3.2}$$

where the number 9 is the ratio of the molecular weight of water (18 g) to the product of the number of hydrogen atoms in a water molecule (2) and the density of water (1 g/cm^3).

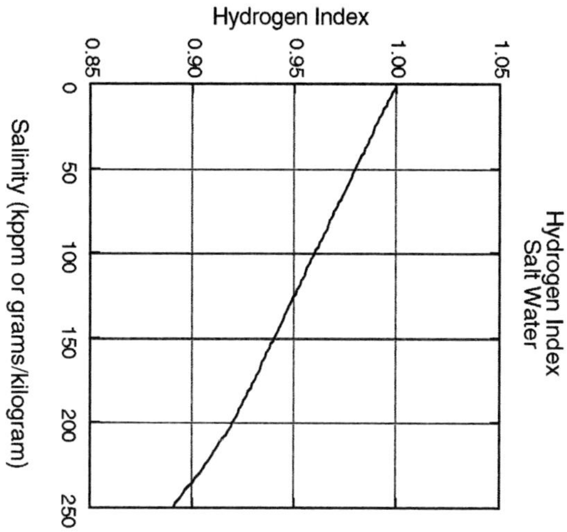

Figure 3.1: Hydrogen index of salt water. (From Kleinberg and Vinegar [128].)

For a light alkane such as decane, the calculation is as follows. Decane has the formula $C_{10}H_{22}$. Thus n_H=22 and the molecular weight is 142. The measured density of decane is 0.7287 g/cm^3. Substituting these numbers in the HI calculation (Eq. (3.2)) yields HI=1.016. Similarly, hexane (C_6H_{14}) with a density of 0.6597 g/cm^3 and a molecular weight of 86 has HI= 0.9665.

Kleinberg and Vinegar [128] reported HI values for a suite of low and medium density oils. Their HI values are plotted versus API gravity in Figure 3.2. For the low density oils (high API gravity), the HI was approximately 1. However, they showed a decrease in HI as the API gravity decreased below 20. They attributed at least some of the decrease in HI to insufficient sampling of the short T_1 and T_2 values of the heavier oils, thus the measured HI may be the apparent HI rather than the true HI.

Note that API gravity[1] of a crude oil is related to its density ρ (in gm/cm^3)

[1]The density of crude oils ranges from about 1.01 to 0.75 gm/cm^3. Thus, the corresponding API gravity varies from about 8 to 57.

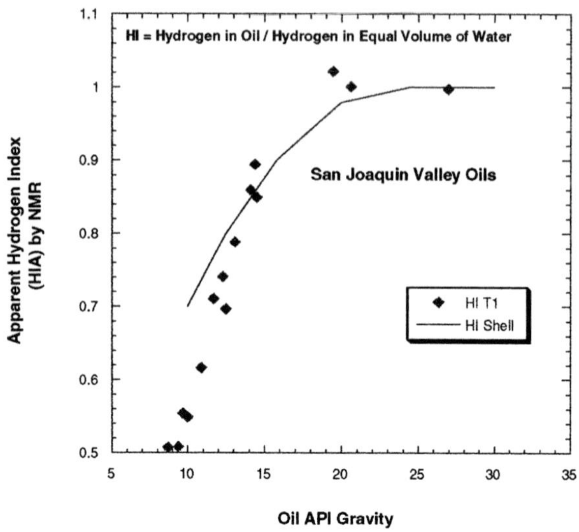

Figure 3.2: Hydrogen Index (HI) versus API gravity. (The solid line is from Kleinberg and Vinegar [128] and the data points are from LaTorraca et al. [137].)

through the following definition:

$$\text{API gravity} \equiv \frac{141.5}{\rho} - 131.5. \tag{3.3}$$

It is a scale (API or American Petroleum Institute scale) commonly used in the petroleum industry to characterize the quality of a crude oil.

LaTorraca et al. [134, 135] reported HI measurements on a suite of medium to high density oils. Their results are also shown in Figure 3.2. The viscosity values of these oils varied from ~1000 cp to ~100,000 cp at room temperature. Both the combustion and low field T_2 methods were used to determine the range in HI. From the combustion method, only the highest density crude oils, with viscosity values greater than ~10,000 cp, had HI values less than one as shown in Figure 3.3. Note that the NMR and combustion methods coincide only for the two oils with the lowest viscosity and density. The HI estimates are plotted versus viscosity because in certain well logging situations it is possible to estimate oil viscosity from the difference between NMR and density porosity determinations.

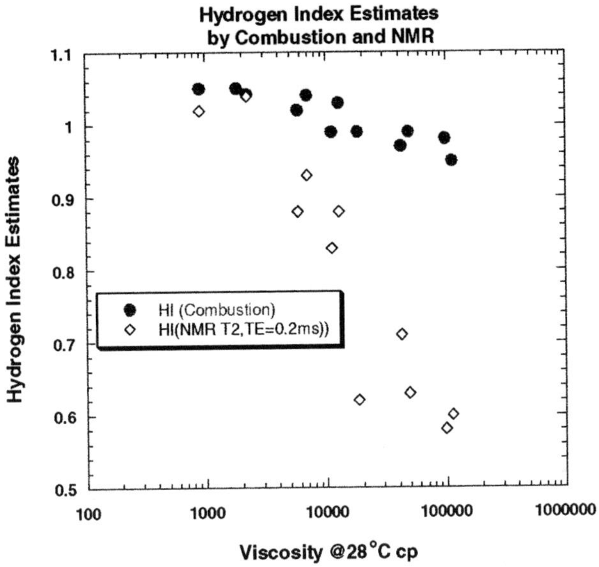

Figure 3.3: Hydrogen Index (HI) determinations on high density oils using combustion and low field NMR methods. (From LaTorraca et al. [135].)

The HI estimates at a Larmor frequency of 2 MHz are plotted in Figure 3.4 for various echo spacings. The apparent HI values depend strongly on the echo spacing. This dependence was explained as being caused by the fast T_2 relaxation rate of high density crude oils being too rapid to be adequately sampled by the low field NMR instruments. This resulted in the low values of apparent HI shown for all the echo spacings used.

The foregoing discussion on HI is mostly for dead oils. For live oils, the dissolved gas can significantly reduce the density such that the HI can be much less than one. Figure 3.5 shows a result for live oils where the HI is between 0.8 and 0.9. The linear lines are HI for different hydrogen:carbon ratios. The curve of G&P is the correlation developed by Gaymard and Poupon [92] based on alkanes as a function of density. It will not describe the HI for heavy oils because heavy oils contain a significant aromatic and polyaromatic content. The curve of K&V is the correlation of HI and the API gravity by Kleinberg and Vinegar [128].

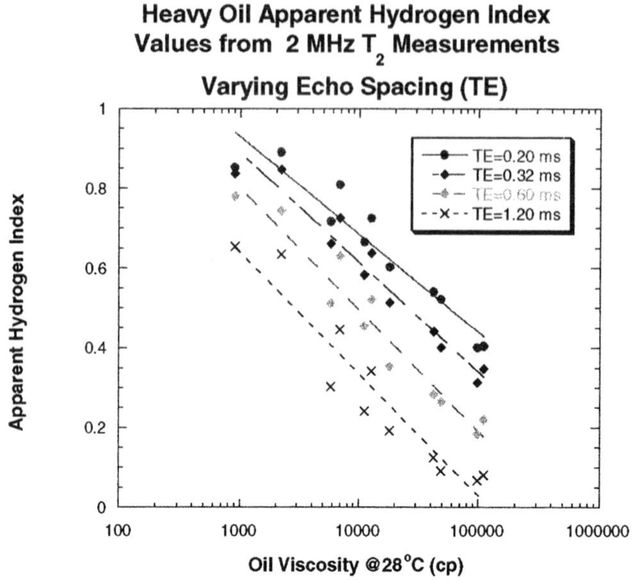

Figure 3.4: Low field NMR HI estimates vs viscosity with varying echo spacing (TE). (From LaTorraca et al. [135].)

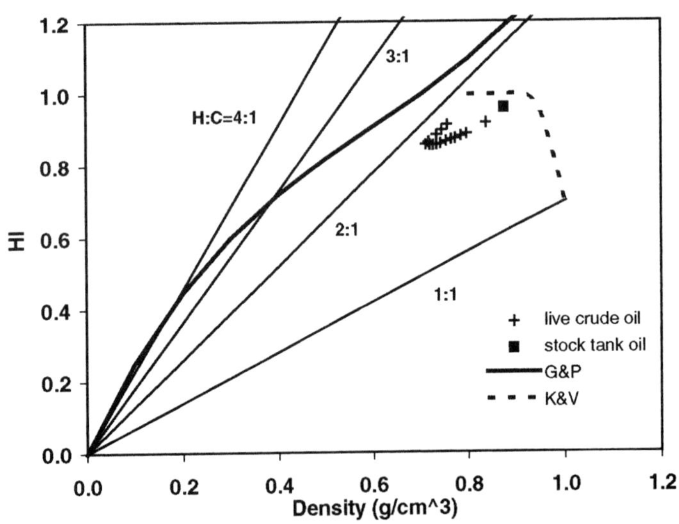

Figure 3.5: HI of live crude oils calculated from PVT data and measurements at ambient condition. (From Zhang et al. [230].)

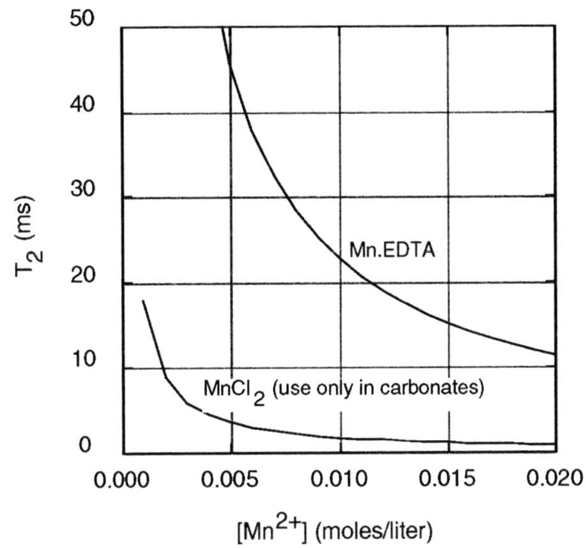

Figure 3.6: The T_2 relaxation time versus the concentrations of Mn^{++} for MnEDTA and $MnCl_2$. (From Horkowitz et al. [103].)

3.2 NMR properties of bulk fluids

3.2.1 Water

At room temperatures and at the low magnetic fields used in NMR well logging, the T_1 and T_2 values of bulk water are in the range of 1 to 3 seconds. The presence of dissolved oxygen, which is paramagnetic, accounts for much of the variation seen in T_1 and T_2 at low magnet strengths. Ferromagnetic and/or paramagnetic ions (such as iron or manganese) can greatly reduce the T_1 and T_2 values of brine. The dependence of T_2 on the concentrations of MnEDTA and $MnCl_2$ is shown in Figure 3.6, and its dependence on the temperature is shown in Figure 3.7. Note that only small concentrations of Mn are needed to drastically reduce the T_2 value.

Because the bulk T_1 and T_2 values for water are so long, they usually contribute very little to the measured relaxation rate of brine saturated porous media. The measured relaxation rate of rocks is normally dominated by surface relaxation which is covered in Chapter 4. The method of doping the borehole fluids or the drilling mud with ferro- or paramagnetic ions, which drastically reduce their T_1 and T_2, is frequently used for residual oil determi-

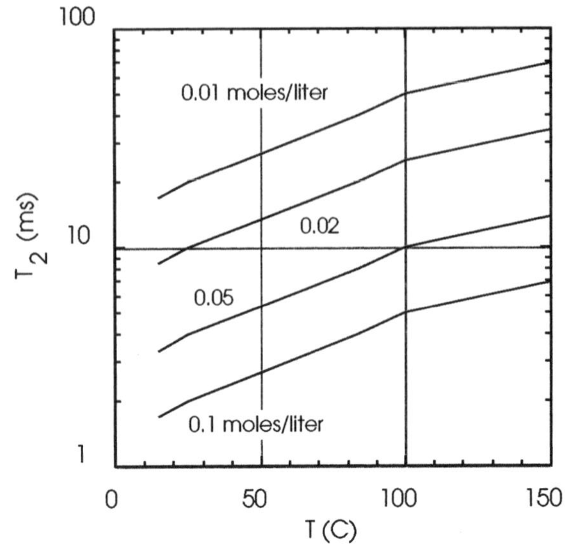

Figure 3.7: The T_2 relaxation time versus temperature for various concentrations of MnEDTA. (From Oakes and Smith [172].)

nation. This is discussed in Chapter 5.

3.2.2 Crude oils

Oil is often the non-wetting phase. It resides in the center of the pore space, and is prevented from making contact with the pore wall. In this situation, there is only bulk relaxation, which is unaffected by the formation and is only a function of oil composition and formation temperature.

However, if the oil wets the surface, the surface relaxation mechanism comes into play just in the same manner as it does for water. The shift of the oil relaxation times when the oil goes from a non-wetting phase to a partially wetting phase can be used to study the wettability issue. This is discussed in Chapter 4. For now, we shall assume that oil is the non-wetting phase, and its NMR relaxation time is the sole property of the oil itself.

For pure alkanes, pure oils, or light crudes, the T_1 or T_2 has a single, well-characterized value. However, for crude oils in general, they are usually characterized by a distribution of relaxation times. In this case, the T_1 or T_2 is represented by multiple amplitudes f_j corresponding to a distribution of relaxation times T_j. There are more discussions on this in Chapters 4 and 6.

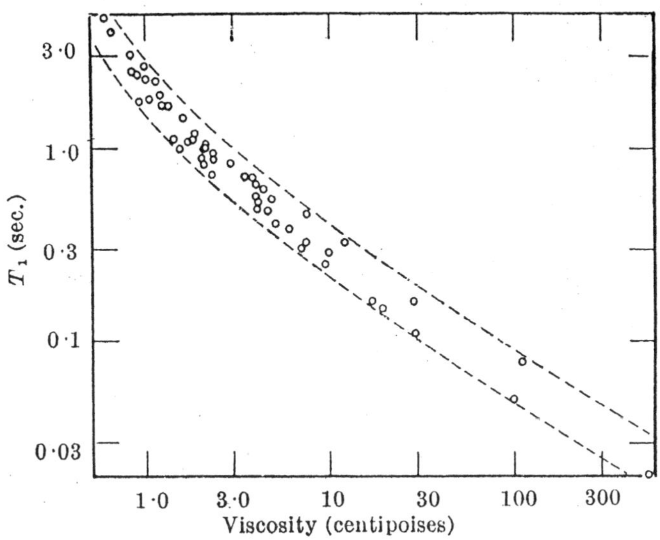

Figure 3.8: T_1 relaxation times versus viscosity for 14 crude oils at various temperatures. (From Brown [40].)

In 1961, Brown [40] studied the T_1 relaxation times for a number of crudes over a wide range of temperatures and viscosities. The relaxation times are obtained from the reciprocals of the initial slopes of the relaxation curves, that is, the reciprocals of the average relaxation rates. His results are shown in Figure 3.8.

Later, Morriss et al. [169] measured the T_2 distributions of a variety of crude oil samples in a uniform magnetic field (Figure 3.9). They found that, in contrast with refined oils which have narrow T_2 distributions, these crude oils have T_2 relaxation times which span several decades.

As shown in Figure 3.9, when the viscosity increases from 2.7 cp at the top of the left hand panel to 4304 cp at the bottom of the right hand panel, the T_2 distribution starts to broaden with the increase of the short relaxation time component, and eventually loses all the long time components and retains only the short time components when the oil is very viscous. This reflects the nature of the crudes which are mixtures of various hydrocarbon types, the mobile ones with small molecular sizes as well as the less mobile ones with long hydrocarbon chains.

Frequently, a single value, such as the logarithmic mean of the T_2 distribution, is used to characterize the overall property of T_2. If the T_2 distribution has a set of multiple relative amplitudes f_j corresponding to a distribution of relaxation times T_j, then the logarithmic mean of the relaxation times, $T_{2,LM}$ is defined as:

$$\log T_{2,LM} = \frac{\sum_j f_j \log T_j}{\sum_j f_j}. \tag{3.4}$$

This logarithmic mean of relaxation times, $T_{2,LM}$, is sometimes referred to as the geometric mean, T_{2G}. Both $T_{2,LM}$ and viscosity for each sample are shown in Figure 3.9.

3.2.3 Gas

For most liquids and solids, the spins relax by dipole-dipole interaction. This relaxation is caused by the interaction of the magnetic dipoles of the nucleus of interest with the dipoles of other nuclei. The relaxation rate is proportional to η/T_K (i.e., the viscosity/absolute temperature). For simple gases, such as methane, the spins relax primarily by spin-rotation interaction. This is the coupling between the magnetic moments of the protons and the magnetic field produced at their positions by the rotation of the molecule. In contrast with fluids and solids, this kind of relaxation rate is proportional to T_K/η.

Unlike liquids, the T_1 of methane has an opposite behavior: It increases as the pressure increases, and decreases as the temperature increases. Figure 3.10 shows the T_1 of methane gas versus temperature and pressure.

A simple power law was found to describe the data shown in Figure 3.10 quite well [93, 182, 178]:

$$T_{1,g} = 2.5 \times 10^4 \frac{\rho}{T_K^{1.17}}. \tag{3.5}$$

where T_K is the absolute temperature in Kelvin, and ρ is the density in g/cm^3.

3.2.4 Mud filtrate

Water based mud

There are various kinds of water based drilling muds. It is always advisable to measure the T_1 and T_2 of the mud filtrate before a logging job. Depending on the degree of invasion of the mud filtrate and the depth of the probed

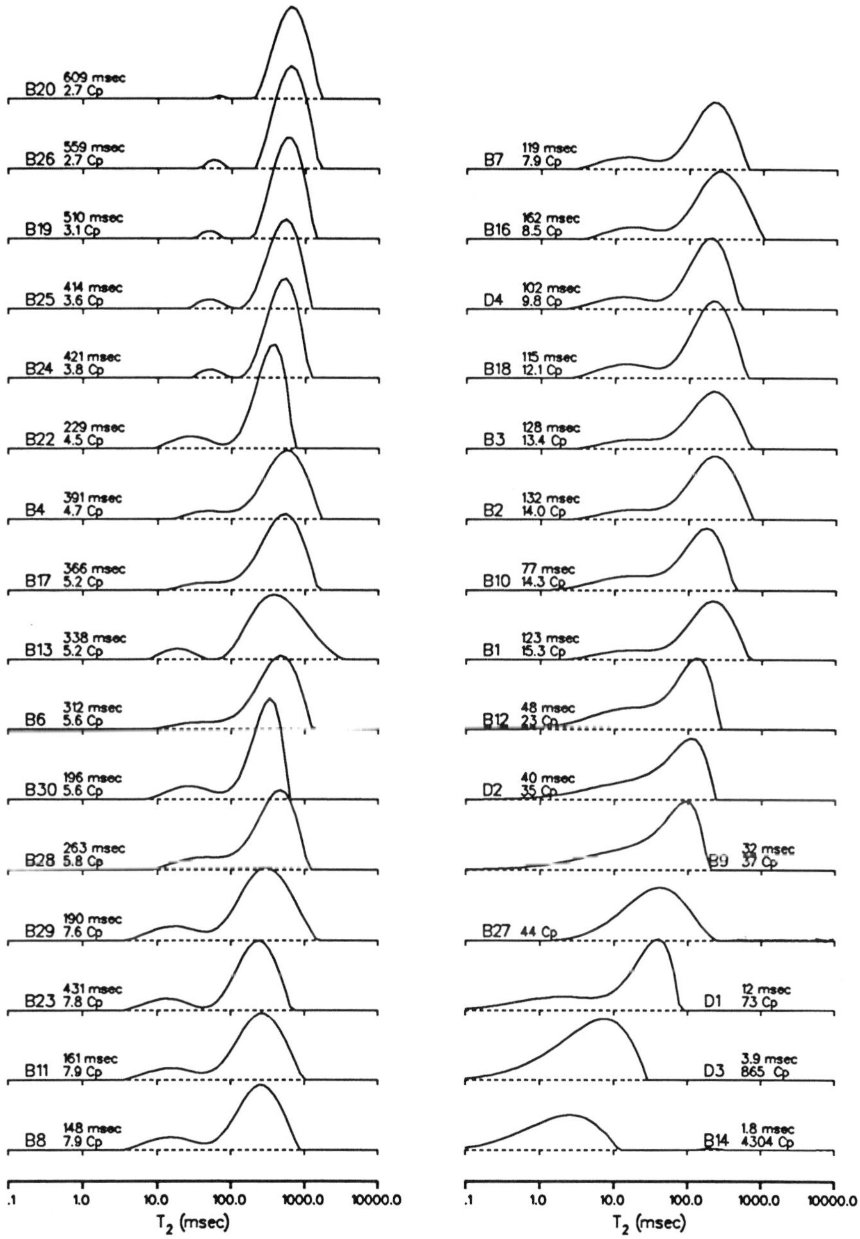

Figure 3.9: The T_2 distributions for bulk oil samples from Belridge field plotted in order of increasing viscosity. The sample number, logarithmic T_2 mean, $T_{2,LM}$, and measured viscosity (in centipoise) are shown for each sample. (From Morriss et al. [169].)

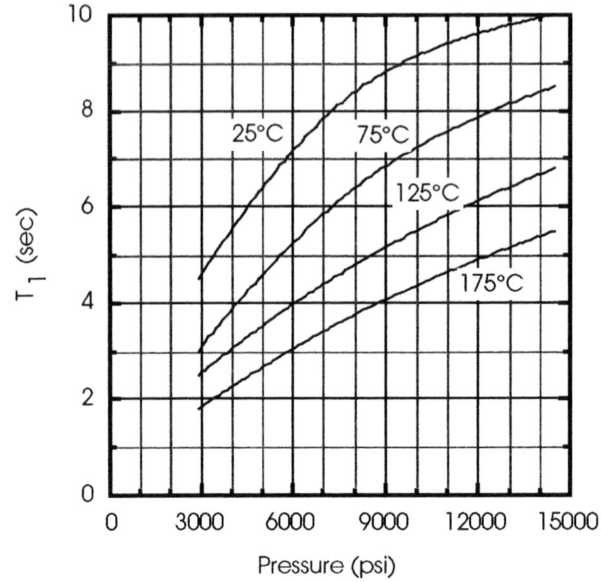

Figure 3.10: T_1 of methane gas versus temperature and pressure. (From Gerritsma et al. [93].)

zone of the logging tools, frequently we are measuring the mud filtrate saturated portion of the formation. In some water-based drilling muds, additives can reduce the relaxation time significantly. An example of such a mud is chromium or ferro-chromium lignosulfonate mud. Figure 3.11 shows T_1 as a function of thinners added to drilling mud from the early work of Coolidge et al. [66].

Oil based mud

Oil based drilling mud (OBM) is used both for environmental reasons and to ensure the stability of the borehole. Oil-based muds typically contain a base oil and at least two additives such as a surfactant and an emulsifier. The purpose of the surfactant is to make the solid particles in the mud oil-wet to block water that is not emulsified from entering the formation. The purpose of the emulsifier is to maintain the water in the mud in an emulsified state. The filtrate from these muds can invade the formation and displace and/or mix with the reservoir fluids. There is no known published study yet on how the various OBM filtrates interact with reservoir fluids and affect the T_2

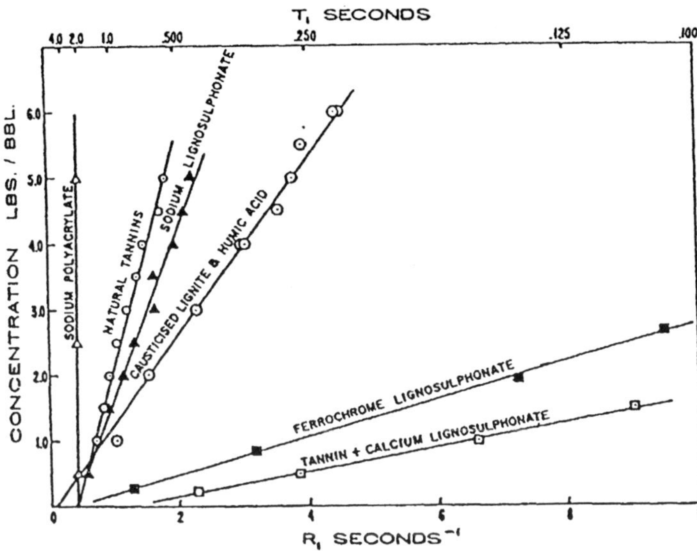

Figure 3.11: Concentration of thinners added to drilling mud versus T_1 relaxation rate, $R_1 \equiv 1/T_1$. (From Coolidge et al. [66].)

measured by NMR logging tools. However, it is known that the presence of these additives in the OBM filtrate can affect the T_2 response.

Figure 3.12 shows the results of the T_2 measurements on the fluid constituents of an oil based mud plus a filtrate sample from the mud. The fluid constituents were the base oil, emulsifier, and surfactant. Note that the T_2 distribution for the base oil is essentially a single component while the surfactant and the emulsifier have complicated, bimodal distributions. Note also that the T_2 for the filtrate is shorter than the base oil T_2 and that a shorter peak indicates that either paramagnetic impurities or some of the surfactant or emulsifier is in the filtrate. Additionally, the surfactant has very short components consistent with its asphaltene (high viscosity, short T_2) constituent.

3.3 T_2 in a magnetic field gradient

For T_2 relaxation in a magnetic field gradient, the dephasing of spins in an infinite medium due to diffusion results in additional signal decay. This leads

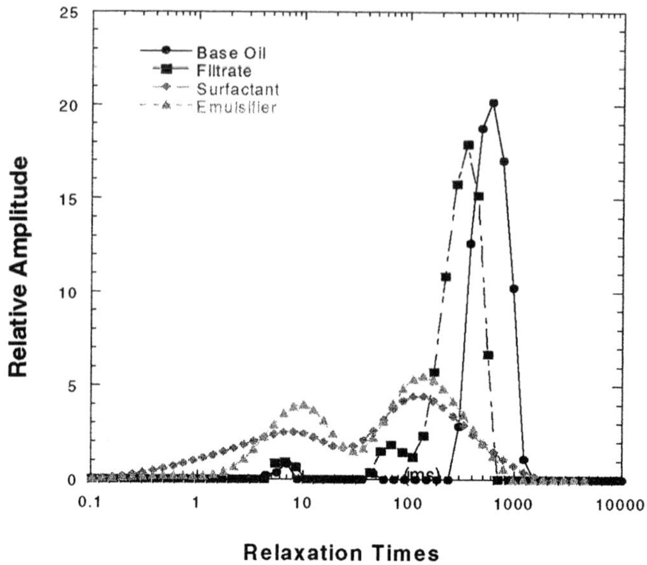

Figure 3.12: T_2 distributions of an oil based mud (containing base oil, surfactant, and emulsifier) and the mud filtrate. (From LaTorraca et al. [136].)

to an additional effective relaxation rate $1/T_{2D}$

$$\frac{1}{T_2} = \frac{1}{T_{2B}} + \frac{1}{T_{2D}}, \tag{3.6}$$

where

$$\frac{1}{T_{2D}} \equiv \frac{(\gamma G t_E)^2 D}{12} \tag{3.7}$$

and γ is the gyromagnetic ratio, G the magnetic field gradient (assumed to be constant), t_E (TE) the time between echoes, D the self diffusion coefficient of the fluid, and T_{2B} the bulk relaxation time (see Chapter 2 for additional details).

Diffusion of spins in a magnetic field can cause incomplete refocusing in a spin echo experiment. This leads to a reduction in signal amplitude in addition to the regular T_2 decay. Since this is an issue of dephasing of spins due to diffusion, it is strictly a T_2 effect (non-existent in T_1)[2]. The net result of this diffusion effect is equivalent to having an additional relaxation rate term as shown in Eqs. (3.6) and (3.7).

[2]See footnote [1] of Chapter 2 on p. 29.

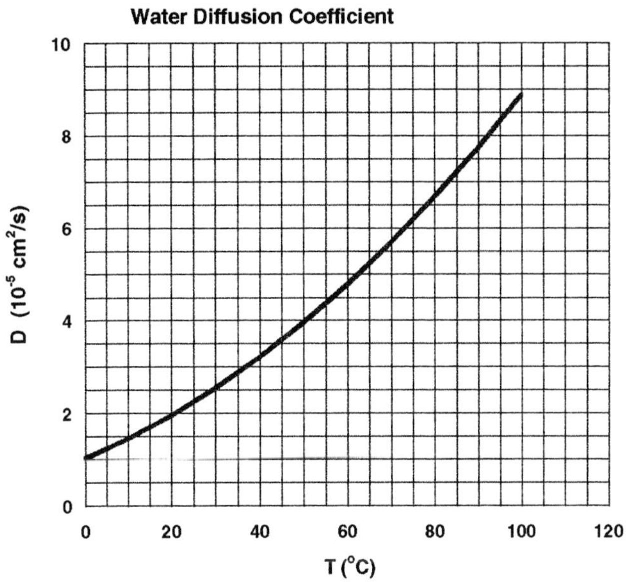

Figure 3.13: Diffusion coefficient of water as a function of temperature[3].

When diffusion is restricted in porous media, the diffusion coefficient in Eq. (3.7) may have to be replaced by the restricted diffusion coefficient for a large TE. There are also other issues as to the validity of Eq. (3.7) which we shall discuss in detail in Chapter 7. For illustrative purposes, we shall use the free diffusion coefficient of the fluid to calculate T_{2D} and describe diffusion effects on the overall relaxation rate.

Looking at the expression of Eq. (3.7), the diffusion coefficient, the magnetic field gradient, and the time between echoes (TE), all play roles in determining T_{2D}.

Water

The diffusion coefficient of water is quite insensitive to pressure, but depends strongly on temperature, as shown in Figure 3.13 [3]. Note that the net effect

[3]Note that the plot is based on $D = a(T_K/\eta)$, where a is determined by $D = 2.0 \times 10^{-5}\text{cm}^2/\text{s}$ at 20°C and the viscosity $\eta = 1.002$ cp. The temperature dependence of η is based on Sengers and Watson [197]. This D vs. T plot can be approximated by the empirical fit: $D = 1.0413 + 0.039828T + 0.00040318T^2$ for the range of 0° to 100°C, where T is in Celsius and D is in $10^{-5}\text{cm}^2/\text{s}$.

of this diffusion term, $1/T_{2D}$, is to shift the T_2 relaxation time at zero field gradient (intrinsic relaxation time) to a shorter apparent relaxation time. Figure 3.14 shows such an effect computed for water using the following input

$$\begin{aligned} \gamma &= 2.6752 \times 10^4 \text{ rad sec}^{-1}\text{gauss}^{-1} \\ D &= 2.5 \times 10^{-5} \text{ cm}^2/\text{sec} \\ \text{TE} &= 1.2 \text{ and } 3.6 \text{ msec} \end{aligned}$$

Note that for a large intrinsic T_2, the change in apparent T_2 value is greater, as illustrated by the bending of the curve, and the effect is more significant for larger field gradient values. For an intrinsic T_2 of 1 sec, at a TE of 3.6 ms, its apparent T_2 is about 340 ms at 10 Gauss/cm and about 115 ms at 20 Gauss/cm; whereas for an intrinsic T_2 of 10 ms, there is hardly any change in its apparent T_2 at the same field gradients.

Oil

Typically, the diffusion coefficient for oil is much less than that of water. Figure 3.15 shows the apparent T_2 for a range of diffusion coefficient values from $D = 1 \times 10^{-6}$ to 2.5×10^{-5} cm^2/sec in a field gradient of 20 Gauss/cm and a TE of 3.6 ms. Most oils have a diffusion coefficient near that of the top curve, whereas water has a diffusion coefficient near that of the bottom curve. Thus, there is very little shift of relaxation time for oils due to the field gradient, whereas the shift for water is quite significant. This is an important property which we use to distinguish oil from water in NMR logs.

Gas

The diffusion coefficient for gas, on the other hand, is much larger than that of water. The bulk (unrestricted) diffusion coefficient of methane is given by the following [173, 99]:

$$D = 8.5 \times 10^{-7} \frac{T_K^{0.9}}{\rho} \tag{3.8}$$

where T_K is the absolute temperature in Kelvin, and ρ is the density in g/cm^3. The density of methane versus temperature is shown in Figure 3.16 [8] and its diffusion coefficient versus temperature is shown in Figure 3.17.

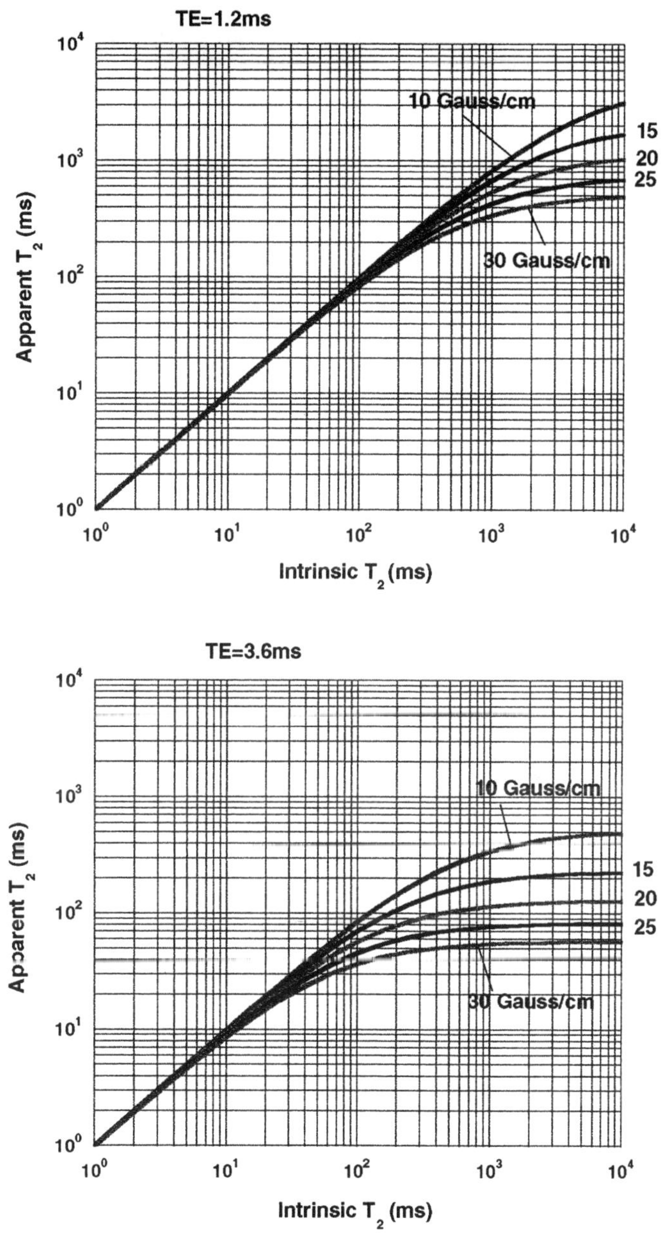

Figure 3.14: The apparent T_2 relaxation time versus intrinsic T_2 relaxation time in a constant magnetic field gradient of various strengths in Gauss/cm and a TE of 1.2 (top) or 3.6 (bottom) ms.

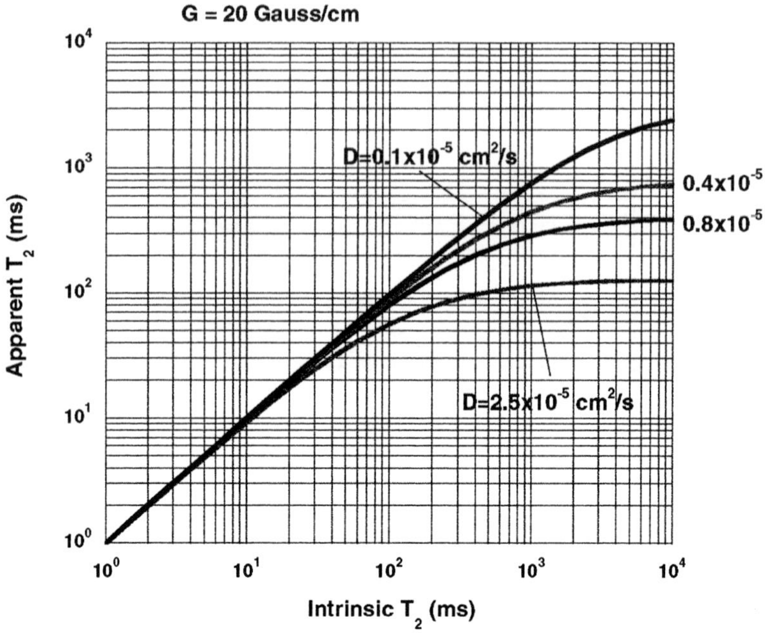

Figure 3.15: The apparent T_2 relaxation time versus intrinsic T_2 relaxation time for four values of diffusion coefficient in a constant magnetic field gradient of 20 Gauss/cm and a TE of 3.6 ms.

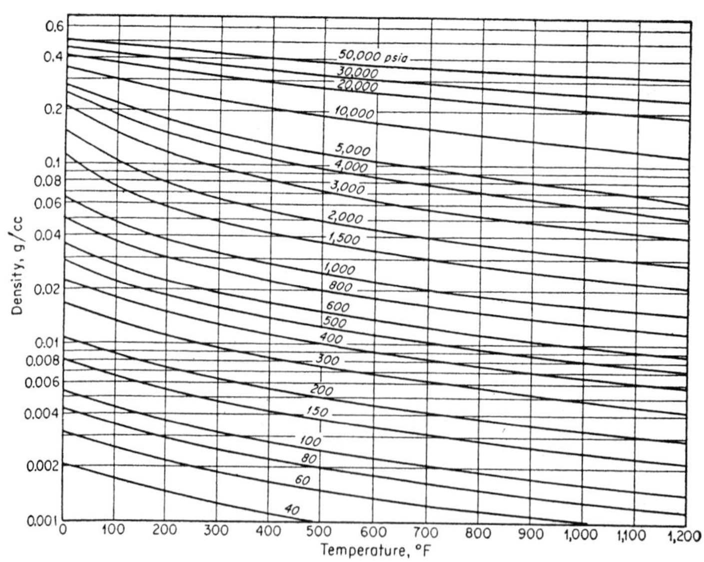

Figure 3.16: The density of methane versus temperature at various pressures. (Reproduced from Fig. 4.17 of Ref. [8] with permission of McGraw-Hill.)

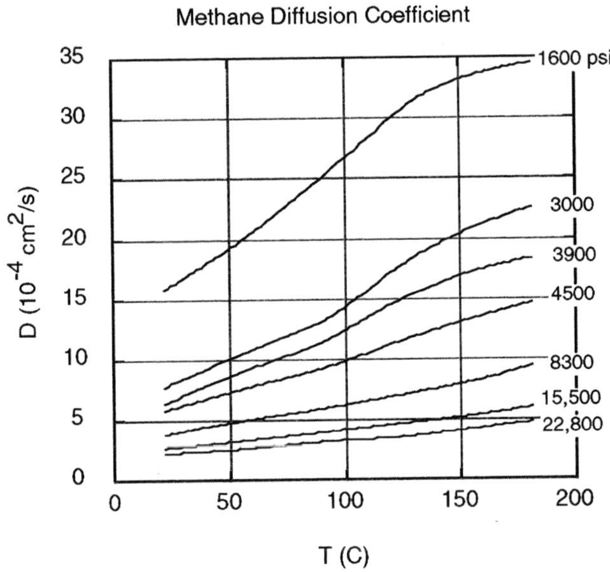

Figure 3.17: The diffusion coefficient of methane versus temperature at various pressures. (From Kleinberg and Vinegar [128].)

A typical value for the diffusion coefficient for methane is of the order of 10^{-3} cm^2/s. Since T_{2B} is of the order of seconds, a value of D of the order of 10^{-4} cm^2/s already makes $1/T_{2D}$ the dominant term on the RHS of Eq. (3.6). Thus, for gas, we have

$$T_2 \approx T_{2D} \equiv \frac{12}{(\gamma G t_E)^2 D}. \tag{3.9}$$

Figure 3.18 shows the apparent T_2 for gas versus its diffusion coefficient in a constant field gradient of 20 Gauss/cm at different TEs. In generating Figure 3.18, we have assumed a T_{2B} for methane of 6 s and used Eq. (3.6) for our calculation. Such an assumption is irrelevant for a TE of 1.2 or 3.6 ms, as the approximation made in Eq. (3.9) is reasonably valid, whereas for a TE of 0.2 ms, a full calculation using Eq. (3.6) is needed.

3.4 Oil viscosity and diffusion constant

One of the important NMR properties of oils is the correlation of the T_1 or T_2 with the viscosity and diffusion coefficient of the oils. This is an important

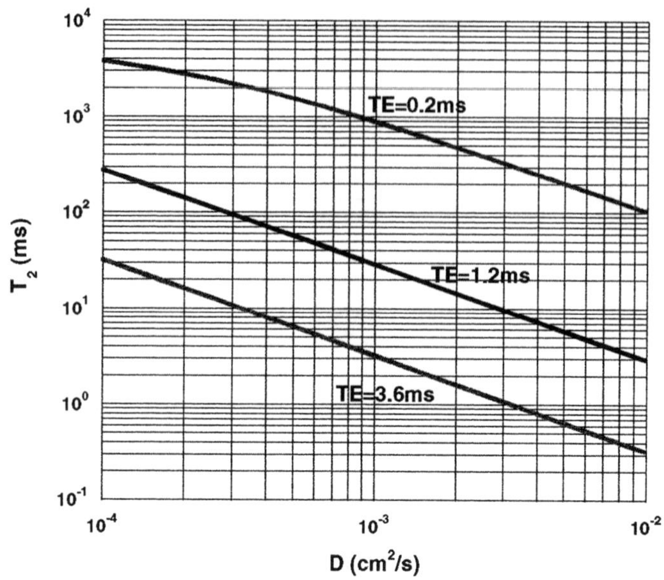

Figure 3.18: Calculated apparent T_2 of methane vs. D for unrestricted diffusion in a field gradient of 20 Gauss/cm.

correlation which is used for the estimation of oil viscosity from the T_2 values obtained from the NMR logs.

3.4.1 Viscosity correlation

Early measurements of T_2 relaxation times of dead crude oils by Vinegar [220] and Morris et al.[169] (Figure 3.19) have established a linear relationship between the logarithmic mean of T_2 and viscosity on a log-log plot

$$T_{2,LM} = \frac{1.2T_K}{298\eta} = 0.00403\frac{T_K}{\eta} \quad \text{(contains dissolved oxygen)} \tag{3.10}$$

where $T_{2,LM}$ is in sec, T_K is the absolute temperature in degree Kelvin, and η is the viscosity in centipoise. The above measurements were done without the deoxygenation procedure, and the samples presumably contained dissolved oxygen of varying degree. The latter is paramagnetic and is known to cause increased relaxation, i.e., shortening of the relaxation time.

Later work by LaTorraca et al. [134, 135] on heavy oils at 2 MHz has shown that the apparent $T_{2,LM}$ may depend on echo spacing. For heavy oils

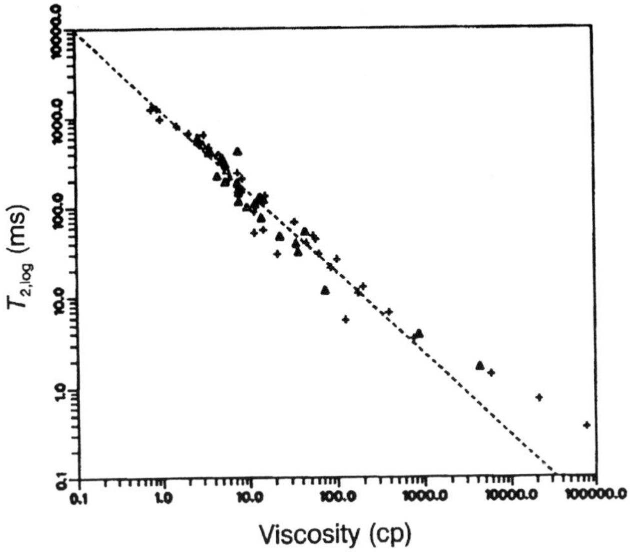

Figure 3.19: $T_{2,LM}$ versus crude oils. The best fit shown in Figure is $T_{2,LM} = 1200/\eta^{0.9}$ where $T_{2,LM}$ is in ms. Later, a more general form as Eq. (3.10) is adopted. (From Morriss et al. [169].)

at large TE, the loss of signal may result in a higher apparent $T_{2,LM}$. They established a formula for correcting this effect:

$$\eta = \left(\frac{2210 + 469\ t_E^2}{T_{2,LM} - (t_E + 0.5)} \right) \left(\frac{T_K}{298} \right) \tag{3.11}$$

where t_E (TE) and $T_{2,LM}$ are in ms. Figures 3.20 and 3.21 show the results of their experiments.

Other measurements by Zega et al. [229] and Zhang et al. [230] on deoxygenated pure alkane and alkane mixtures also established a linear relationship between $T_{1,2}$ and viscosity on a log-log plot[4]

$$T_{1,2} = 0.00713 \frac{T_K}{\eta} \quad \text{(oxygen free)}, \tag{3.12}$$

and later work by Lo et al. [143, 144] has revised the fitting constant:

$$T_{1,2} = 0.009558 \frac{T_K}{\eta} \quad \text{(oxygen free)}, \tag{3.13}$$

[4]Note that Eqs. (3.10), (3.12), and (3.13) are all based upon room temperature measurements. Thus, the T_K dependence is based on theory.

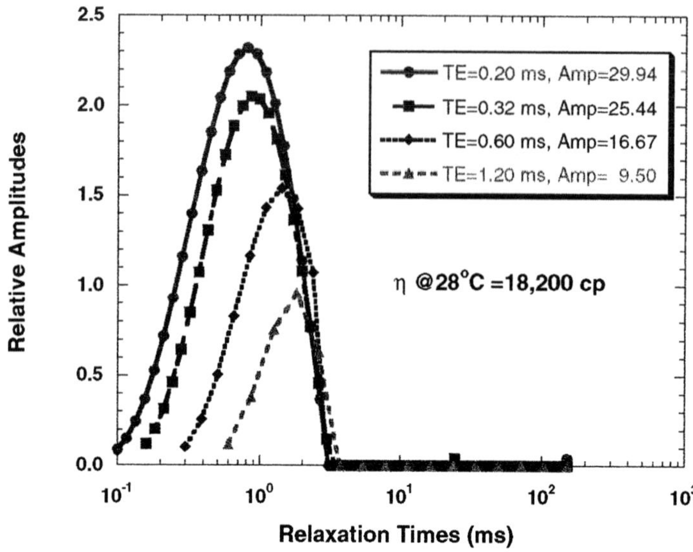

Figure 3.20: T_2 distributions of a heavy oil with varying echo spacing measured at 2 MHz. Note reduction of amplitude and apparent shift of time constants to longer times with increasing TE. (From LaTorraca et al. [135].)

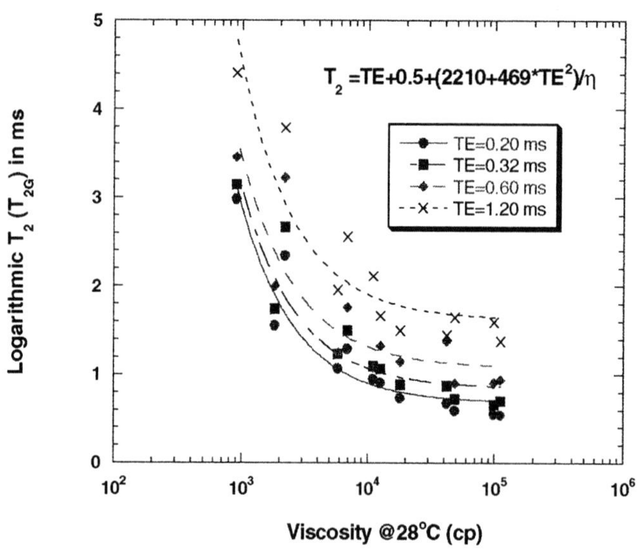

Figure 3.21: Logarithmic mean of T_2 of heavy oils versus viscosity at 2 MHz for varying TE. (From LaTorraca et al. [135].)

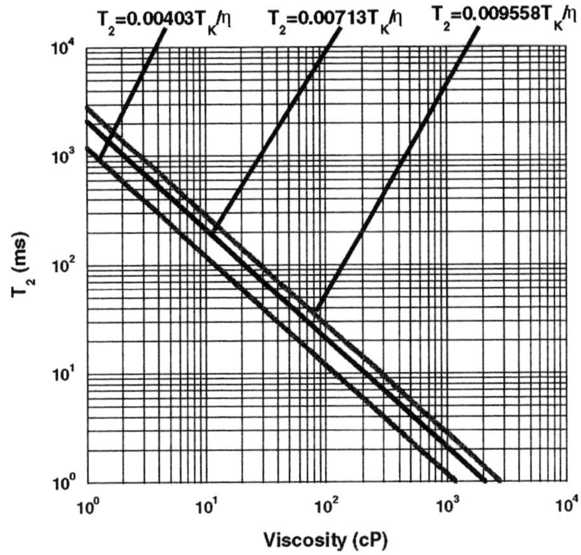

Figure 3.22: $T_{1,2}$ versus viscosity for Eqs. (3.10), (3.12), and (3.13) at room temperature.

where $T_{1,2}$ is in seconds, T_K is the absolute temperature in degree Kelvin, and η is the viscosity in centipoise. The deoxygenation procedure yields a larger $T_{1,2}$ relaxation time, consistent with the hypothesis that the earlier measurements [i.e., Eq. (3.10)] may have dissolved oxygen in the samples. The relationships between $T_{1,2}$ and viscosity of Eqs. (3.10), (3.12), and (3.13) are plotted in Figure 3.22.

3.4.2 Diffusion constants of oils

Using the Pulsed Field Gradient Spin Echo (PFGSE) method (Figure 2.9), Lo et al. [143] measured the diffusion coefficient of a liquid methane-decane mixture. The measured PFGSE amplitude is expressed as follows:

$$M(2\tau)/M(0) = \sum_i f_i e^{-\gamma^2\delta^2 g^2(\Delta - \frac{1}{3}\delta)D_i} \tag{3.14}$$

where the values of Δ (diffusion time), g (pulsed field gradient), and δ (pulse width) can be varied. They assumed a distribution of diffusion coefficients D_i and $\sum_i f_i = 1$, and obtained a a distribution of self diffusion coefficients with

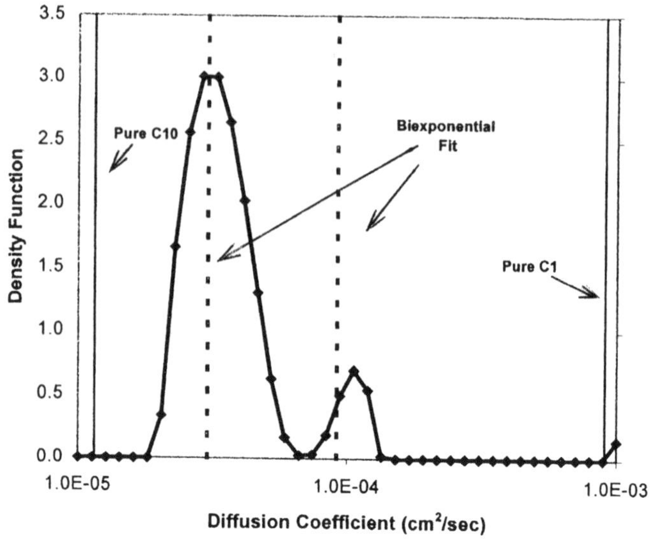

Figure 3.23: Diffusion coefficient distribution of a liquid methane-decane mixture. The proton fraction of methane is 0.187, and that of decane is 0.813; whereas the relative amplitudes in the distribution are 0.189 and 0.811, respectively, quite consistent with the proton fractions. (From Lo et al. [143].)

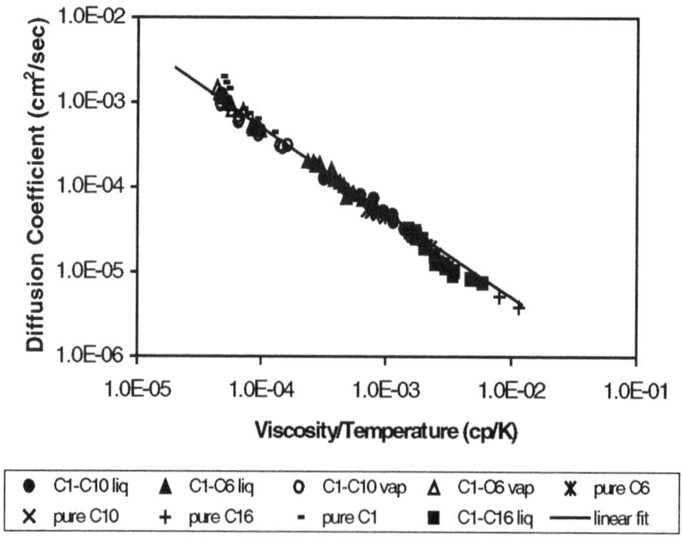

Figure 3.24: Diffusion coefficient dependence on viscosity/temperature for pure methane, pure alkanes, and methane-alkane mixtures. The dependence is linear regardless of composition. (From Lo et al. [144].)

varying relative amplitudes. Their results (shown in Figure 3.23) indicate that there are two components in the diffusion coefficient distribution, each is different from their values in the pure state, while their amplitudes are quite consistent with the relative mole fractions of the constituents.

Further work by Lo et al. [144] shows that the logarithmic mean of the diffusion coefficient distribution of a number of methane-alkane mixtures varies linearly with viscosity/temperature on a log-log plot, regardless of composition. Their results are shown in Figure 3.24. The relation they found can be described by the following equation:

$$D = 5.05 \times 10^{-8} \frac{T_K}{\eta} \tag{3.15}$$

where D is in cm^2/sec, T_K is the absolute temperature in degree Kelvin, and η is the viscosity in centipoise (cp).

3.4.3 Correction for live oils

The linear relationships on a log-log plot between $T_{2,LM}$ and η/T_K, as discussed earlier, are for dead oils. Lo et al. [144] found that the relationship for live oils is no longer linear. Based on experimental measurements on the T_1 relaxation time for pure methane, pure alkanes, and three methane-alkane mixtures (methane-hexane, methane-decane, and methane-hexadecane), they found the T_1 dependence on viscosity/temperature departs from linear correlation, as shown in Figure 3.25.

The T_1 dependence on diffusion coefficient for the systems they studied is plotted in Figure 3.26 for various gas oil ratios (GOR). The GOR is defined as:

$$\text{GOR} = \frac{\text{cubic meters of methane at standard condition}}{\text{cubic meters of stock tank oil}} \tag{3.16}$$

where the standard condition is 60°F and 1 atmospheric pressure.

Since T_1 versus T_K/η and T_1 versus D share the same GORs, the horizontal axis can be either D in cm^2/sec or $5.05 \times 10^{-8} T_K/\eta$.

Figure 3.27 shows the deviation from the linear relationship in the log scale, which is plotted vs. GOR, and a quadratic functional fit determined as follows:

$$\begin{aligned} \text{deviation} &= \log_{10} T_{1,linear} - \log_{10} T_1 = \log_{10}(f(\text{GOR})) \\ &= -0.127(\log_{10}\text{GOR})^2 + 1.25\log_{10}\text{GOR} - 2.80 \end{aligned} \tag{3.17}$$

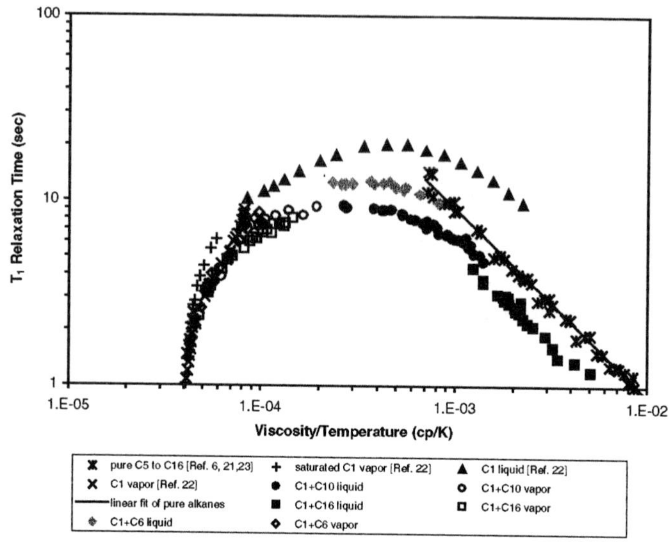

Figure 3.25: T_1 relaxation time for pure methane, pure alkanes, and three methane-alkane mixtures. Its dependence on viscosity/temperature departs from linear correlation. (From Lo et al. [144].)

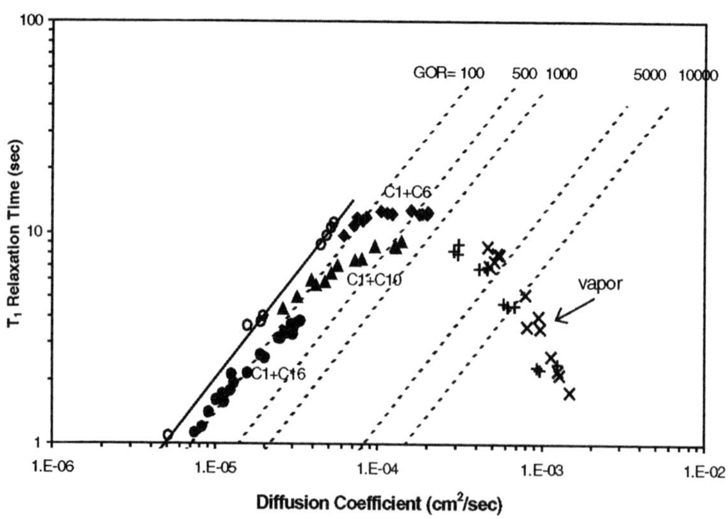

Figure 3.26: T_1 dependence on diffusion coefficient for pure methane, pure alkanes, and methane-alkane mixtures. Contour lines of constant GOR are plotted parallel to zero GOR. (From Lo et al. [144].)

where $T_{1,linear}$ is given by the earlier equations. So, once the GOR and the T_1 of the oil are known, the viscosity can be estimated as follows:

$$T_1 = \frac{0.009558}{f(\mathrm{GOR})}\left(\frac{T_K}{\eta}\right) \tag{3.18}$$

where Eq. (3.13) is used for the linear relation between T_1 and T_K/η.

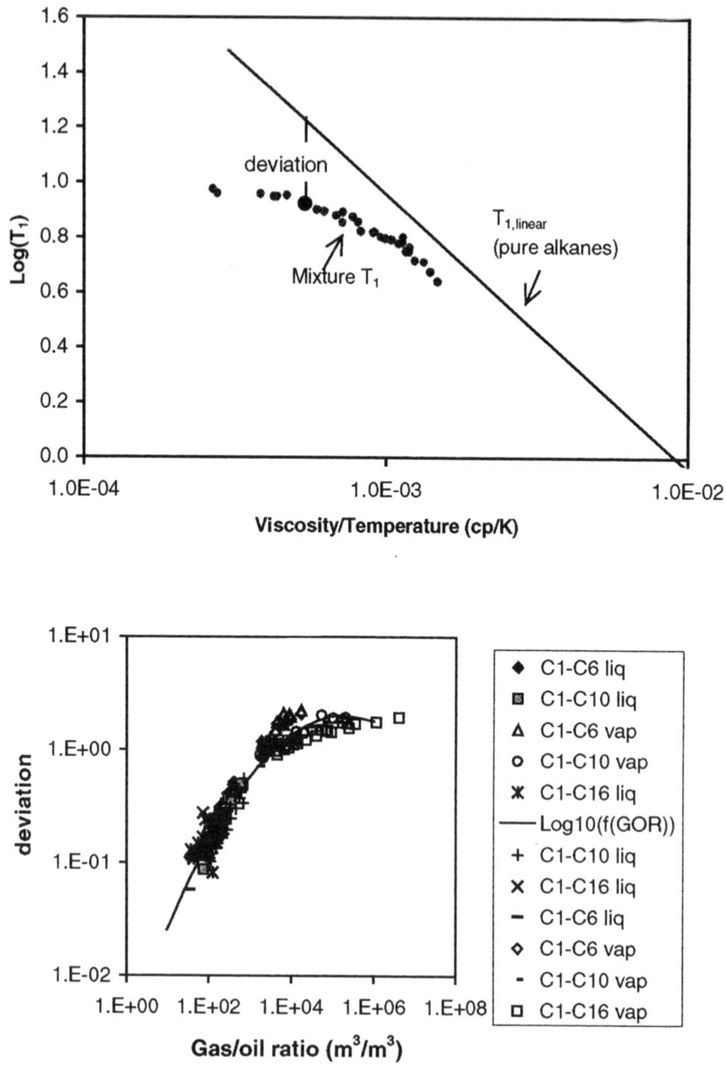

Figure 3.27: The deviation from the linear relationship in the log scale (top) was plotted against GOR (bottom), and a quadratic functional fit was determined and used for correction for live oils. (From Lo et el. [144].)

Chapter 4

Petrophysical NMR measurements

4.1 NMR in porous media

As described in earlier chapters, the nuclear magnetic relaxation behavior of water in porous media tends to be quite different from that of water in the bulk phase. There were experiments [41, 37] performed which demonstrated that water, when it resides in the pore space of a rock (e.g. Berea sandstone), has a much shorter relaxation time (T_1 typically varies from 10 to several hundred milliseconds) than it has in the bulk phase (about 3 seconds). The very same water, when removed from the rock, regains its long relaxation time. Hence, the water that came into contact with the Berea sandstone did not change, or at least did not change significantly enough to have any discernible change in the relaxation time. Accordingly, one can only interpret the fast relaxation of water in Berea sandstones as the result of enhanced surface relaxation at the pore-matrix interface, which is presumably due to a combination of paramagnetic impurity ions and reduced correlation times at the interface.

In porous rocks, we mostly study the relaxation of hydrogen nuclei in the pore space. The values of T_1 and T_2 are affected by the physical and chemical properties of the rock fluids, the uniformity of the the static magnetic field, the variations in pore sizes, the differences between the magnetic susceptibility of pore fluids and rock grains, the fluid viscosity and diffusion, and the presence of paramagnetic impurity ions on the pore walls or in the fluids.

4.1.1 Surface relaxation

The basic physics of NMR relaxation of fluids at solid surfaces in porous media were first studied by Korringa, Seevers, and Torrey [129]. They identified two surface relaxation mechanisms: one occurs at all sites on the surface, and the other is associated with dilute paramagnetic ion impurities. The relaxation process associated with all non-magnetic sites on the surface is much too weak to account for the fast relaxation observed for water in rocks (Almagor and Belfort [6], D'Orazio et al. [73]). Para- and ferromagnetic ions, such as iron, manganese, nickel, and chromium can cause strong surface relaxation. Their presence on the solid surfaces of the rock will dominate the relaxation process of water in the pore space. Typical sandstones have about 1% iron content, and some contain a significant concentration of manganese. This level of paramagnetic impurity ions will make spins relax very efficiently. Carbonates, on the other hand, have more varied mineralogies, large vuggy pores, and a wider range of surface relaxation strength (typically smaller than sandstones). Hence, their relaxation times can sometimes be as long as the bulk water values.

Fast, intermediate, and slow diffusion regimes

There is evidence that the relaxation of polarized proton spins in fluids, which occupy porous rocks or biological tissues, cannot be described by a single relaxation time. Brownstein and Tarr [45] demonstrated that multiexponential decay arises simply as a consequence of an eigenvalue problem of spin diffusion in porous media of various sizes and shapes, even in the case of a very simple shape of an isolated void or biological cell. They considered two properties which govern the evolution of spin polarization in the system: (1) paramagnetic materials on the solid/fluid interface acting as relaxation centers, and (2) classical diffusion of water molecules with the diffusion constant characteristic of bulk water.

The governing equations for diffusion of spins in the porous media can then be written as

$$\begin{aligned} D\nabla^2 m(\mathbf{r},t) - \frac{m(\mathbf{r},t)}{T_B} &= \frac{\partial m(\mathbf{r},t)}{\partial t} \quad \text{in the pore space} \\ \hat{\mathbf{n}} \cdot D\nabla m(\mathbf{r},t) + \rho m(\mathbf{r},t) \Big|_S &= 0 \quad \text{at the interface} \end{aligned} \tag{4.1}$$

where $m(\mathbf{r},t)$ is the density of unrelaxed spins, $1/T_B$ is the bulk relaxation rate, ρ is the relaxation rate at the solid/fluid interface, or surface relaxation strength, D is the diffusion coefficient of spins in the pore fluid, and $\hat{\mathbf{n}}$ is the unit outward normal on that interface (i.e., it points out of the pore space and into the solid matrix). Brownstein and Tarr showed that the magnetization

$$M(t) = \int m(\mathbf{r}, t) d^3\mathbf{r} \tag{4.2}$$

is given by a sum of decreasing exponential functions of time

$$M(t) = M(0) \sum_{n=0}^{\infty} f_n e^{-\lambda_n t} = M(0) \sum_{n=0}^{\infty} f_n e^{-t/T_n} \tag{4.3}$$

where f_n are the relative amplitudes for each exponential decay, and λ_n (or $1/T_n$) are the eigenvalues of the time independent version of Eq. (4.1), i.e.,

$$D\nabla^2 \psi_n(\mathbf{r}) + \lambda_n \psi_n(\mathbf{r}) = 0 \tag{4.4}$$

with the boundary condition

$$\hat{\mathbf{n}} \cdot D\nabla \psi_n(\mathbf{r}) + \rho \psi_n(\mathbf{r}) \Big|_S = 0 \tag{4.5}$$

where the bulk relaxation e^{-t/T_B} has been factored out. (Note that the bulk relaxation term is often neglected in the discussion because typically $T_B \gg T_n$ and the relaxation is dominated by $1/T_n$. However, this is not always true as we shall discuss below.) These eigenvalues, λ_n $(1/T_n)$, are primarily determined by the geometry of the pore space. However, their values depend strongly on ρ, the surface relaxivity, and D, the diffusion constant of water.

Brownstein and Tarr found that when $\rho a/D \ll 1$, where a is a characteristic dimension of the system they considered, the lowest eigenvalue, $1/T_0$, is given by

$$\frac{1}{T_0} \cong \rho \frac{S}{V} \equiv \frac{\rho}{a} \tag{4.6}$$

where S is the pore surface and V is the pore volume. a can now be alternatively defined as V/S, a characteristic dimension of the porous system. Thus, the condition $\rho a/D \ll 1$ occurs when the the time it takes for the spin to diffuse to the solid/fluid interface, $t \sim a^2/D$, is much shorter than the spin relaxation time, a/ρ, i.e., a fast diffusion situation.

Based upon the three simple pore shapes which they studied, namely parallel slab, circular cylinder, and sphere, Brownstein and Tarr categorized the diffusion into three different regimes, i.e.,

$$\begin{array}{lrl} & \textit{fast}- & \rho a/D \ll 1, \\ & \textit{intermediate}- & 1 \ll \rho a/D \ll 10, \\ \text{and} & \textit{slow}- & 10 \ll \rho a/D \end{array}$$

diffusion regimes.

In the *fast* diffusion regime, there is a large gap between the lowest eigenvalue, $\rho S/V$, and the second lowest and higher eigenvalues. Thus, the lowest eigenvalue is the dominant term, and it depends only on S/V: No other detail of the microstructure enters. This is because the fast diffusion equalizes the spin density everywhere within the pore space. The fine details of the pore geometry thus do not play a role in the magnetization decay, except at extremely short times.

In the *slow* diffusion regime, the lowest eigenvalue is of the order of a^2/D, and expectedly, somewhat different for different pore shapes. Its value is very close to the higher eigenvalues which contain detailed information on the pore geometry. In this case, the magnetization decay has a multiexponential behavior, even for a single pore.

Figure 4.1 illustrates the fast and slow diffusion behavior in a simple pictorial way. On the left is the *fast diffusion regime.* Because of the fast diffusion and the small surface relaxivity, the unrelaxed spin density is uniform over the pore space. When these spins wander around and come into contact with the surface and relax, the unrelaxed spin population is quickly equalized through the diffusion process. As a result, the magnetization decays with a uniform spin density. On the right is the *slow diffusion regime.* In this case, the surface relaxivity is large and the diffusion is slow. The unrelaxed spin density near the surface decreases faster than the rate at which the unrelaxed spins can flow in, by diffusion from other parts of the pore space, to replace the loss. The spin density thus develops a profile which is high at the center of the pore and near zero at the pore surface.

The temperature and pressure effects

If rocks are in the slow diffusion regime, then their relaxation rates should depend on the self diffusion coefficient of water which is very sensitive to

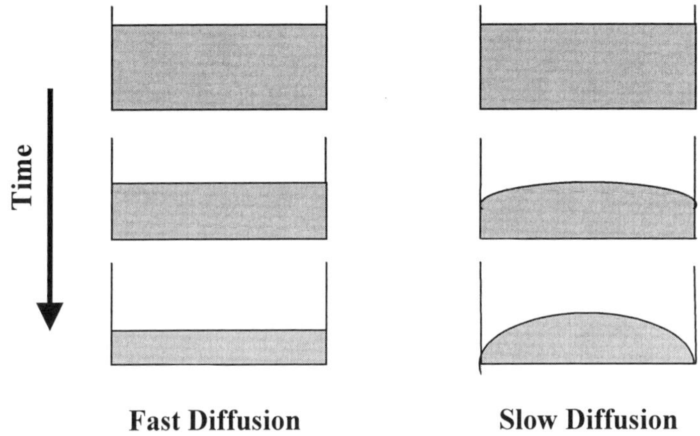

Figure 4.1: Schematic diagram of the slow and fast diffusion regimes

temperature changes. Latour, Kleinberg, and Sezginer [138] have demonstrated experimentally that T_1 and T_2 of water-saturated sandstones and some carbonates are only weakly dependent on temperature over the range of 25° to 175°C. Figure 4.2 shows some of their results. The lack of temperature dependence of relaxation times has been taken as evidence that water-saturated rocks are in the *fast diffusion regime.* It also suggests that the surface relaxivity is independent of temperature.

In the same paper, Latour et al. [138] also demonstrated that the T_1 relaxation behavior of oil-saturated rocks cannot be properly explained by the slow diffusion analysis. If the fast diffusion regime is assumed, they found $\rho_{oil} = \rho_w/17$ in one case, and $\rho_{oil} = \rho_w/30$ in another. However, there is a significant difference between the temperature behavior of the T_1 distributions of water-saturated rocks versus that of oil-saturated rocks. For water-saturated rocks, the surface relaxation plays a dominant role, and the bulk relaxation term can practically be neglected. Hence, if the surface relaxivity is temperature independent, so is the T_1 distribution. This is not true in the oil-saturated rocks. Since $\rho_{oil} \ll \rho_w$, the surface relaxation is very weak, and the oil bulk relaxation, which is temperature dependent, cannot be neglected and has to be included in the consideration. As a result, even

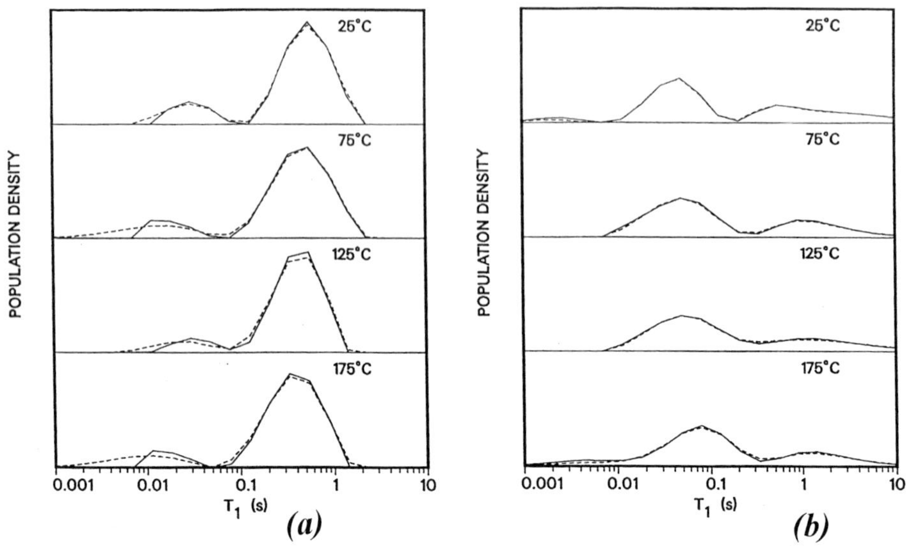

Figure 4.2: The T_1 distributions of water-saturated sandstones (a) and carbonates (b) at four different temperatures. (From Latour et al. [138].)

though the surface relaxation is in the fast diffusion regime and independent of temperature, the overall T_1 distributions, when including the temperature sensitive bulk relaxation, can be quite sensitive to temperature changes.

Recent work on carbonates by Ramakrishnan et al. [183, 184] and Fordham et al. [86] indicated that the carbonates which Latour et al. [138] studied are only a subset of carbonates whose T_1 and T_2 distributions are weakly temperature dependent. They found that although that the T_2 distributions of most sandstones and some carbonates that Latour et al. studied are weakly temperature dependent, there are other types of carbonates whose T_2 distributions show strong temperature dependency. Figure 4.3 shows one such type of carbonate sample which displays a strong temperature dependence in its T_2 distributions. They suggested a model which explains the temperature dependent behavior of the T_2 distributions of some types of carbonates, but fails to explain the others (see discussion on carbonates in Section 4.1.3).

Chen et al. [57] studied the hydrostatic pressure effect on proton and deuteron magnetic relaxation behavior of water in rocks and glass bead packs. From ambient to 250 MPa, they found very little change in the T_1 distributions of H_2O and D_2O saturated porous systems. The weak pressure depen-

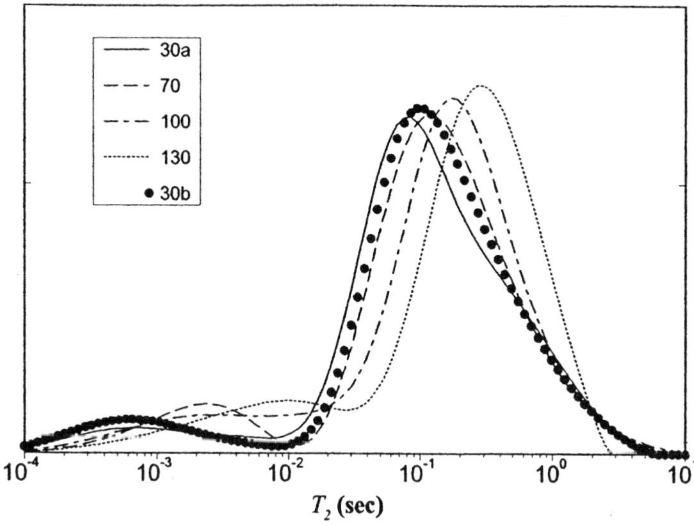

Figure 4.3: The T_2 distributions from 30°C to 130°C for a carbonate rock sample. (From Ramakrishnan et al. [184].)

dence of D_2O relaxation in porous media appears to follow that of the bulk liquid.

Relaxation of water-saturated rocks

Now that we know the relaxation of water-saturated rocks is usually in the *fast diffusion regime*, the total relaxation rate of water in the pore space is a simple volumetric rate average without involving the diffusion constant. Suppose there is a surface layer of thickness λ, such that any spin which enters this layer relaxes at the surface relaxation rate $1/T_{1m}$, where the subscript m stands for mineral or matrix. The total relaxation rate of the water in the pore space $1/T_1$ can be written as:

$$\frac{1}{T_1} = \frac{1}{T_{1B}}(1 - \frac{\lambda S}{V}) + \frac{1}{T_{1m}}\frac{\lambda S}{V}, \tag{4.7}$$

where S and V are the surface and volume of the pore space, respectively. The bulk relaxation time, T_{1B}, is of the order of 1 second, and the surface relaxation time, T_{1m}, for water in sandstones, is of the order of 0.1 to 1 millisecond. If the second term on the RHS of Eq. (4.7) is much larger than the first term, we can write:

$$\frac{1}{T_1} \approx \frac{\rho S}{V} \equiv \frac{1}{T_{1S}}, \qquad \rho \equiv \frac{\lambda}{T_{1m}}, \tag{4.8}$$

where $1/T_{1S}$ is the surface relaxation rate. We now see that the total relaxation is dominated by the surface relaxation, and the relaxation time T_{1S} is proportional to the pore size a, which is characterized by V/S. Since the spins diffuse very fast in the pore space, the earlier implicit assumption of paramagnetic impurity ions being uniformly distributed over the pore surface is really irrelevant. All spins within a pore will simply relax at a monoexponential rate given by $\exp(-t/T_{1S})$. Similar arguments and Eqs. (4.7) and (4.8) also hold true for T_2.

The self diffusion coefficient for water at 25°C is: $D = 2.5 \times 10^{-5}\text{cm}^2/\text{sec}$. For one second, a spin can travel a distance of the order of $\sqrt{6Dt}$, which is about 120 μm. For sandstones, this is typically a distance of a few pores. Travelling through a narrow pore throat, the spin is more likely to relax because of higher frequency of hitting the surface. As a result, the signal of the total magnetization of the rock is the sum of all relaxation rates, each of which is associated with an individual pore, or at most a group of neighboring pores. Since, each relaxation time is proportional to the pore size, this becomes the basis for the assertion that NMR relaxation time distribution reflects the pore size distribution.

4.1.2 Pore size distribution

Based on the understanding in the previous section, we can describe the relaxation of the magnetization of a rock as follows:

$$M(t)/M_0 = \sum_{j=1}^{n} f_j e^{-t/T_j}, \qquad \sum_{j=1}^{n} f_j = 1, \tag{4.9}$$

where $M(t)$ and M_0 represent the magnetization at time t and 0, respectively, and f_j the volume fraction of those pores which have a relaxation time T_j. Each relaxation time, $T_j = a_j/\rho$, is characterized by its own pore dimension a_j and decoupled from other T_j's which have other values of a_j. That is, each individual pore or a group of neighboring pores, is relaxing more or less independently of other pores because of the short diffusion distance we discussed above.

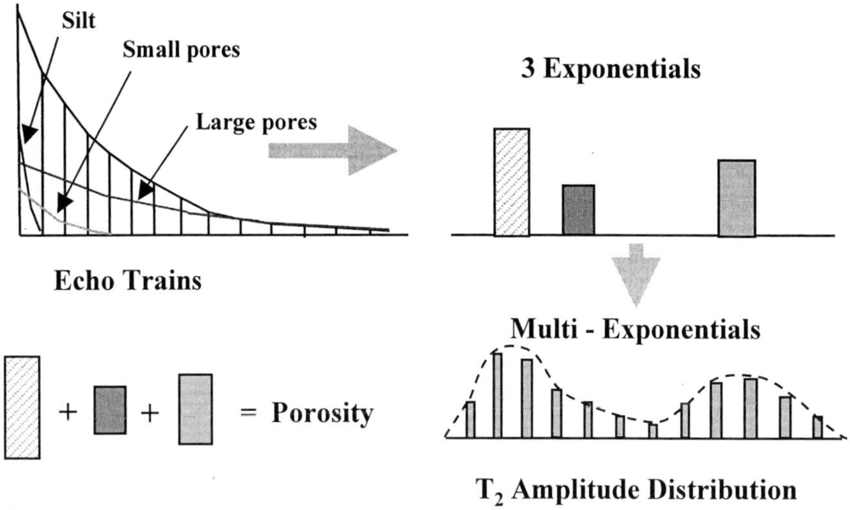

Figure 4.4: A typical analysis of water-saturated rocks with three exponential decaying curves from a nonlinear fit, as well as a continuous distribution of relaxation times equally spaced on a logarithmic scale with corresponding amplitudes from a linear inversion process.

To illustrate this, consider a possible time decaying NMR signal from a brine-saturated rock as shown in Figure 4.4. The total signal is the sum of those signals from the fluid in silt or shale, small pores and large pores, each decaying at a monoexponential rate with a relaxation time T_j which is proportional to the pore size in those regions, and with an amplitude f_j which is proportional to the amount of hydrogen nuclei (or volume fraction) in those regions.

There are several methods for obtaining the values of volume fractions f_j and relaxation times T_j. An early approach [212, 213] assumed a two- or three-exponential model as follows:

$$\begin{aligned} M(t)/M_0 &= f_A e^{-t/T_A} + f_B e^{-t/T_B} \\ 1 &= f_A + f_B, \end{aligned} \tag{4.10}$$

or

$$M(t)/M_0 = f_A e^{-t/T_A} + f_B e^{-t/T_B} + f_C e^{-t/T_C}$$

$$1 = f_A + f_B + f_C, \tag{4.11}$$

To analyze a set of data, for example $M(t_i)/M_0$, we try to minimize the following quantity

$$\chi^2 = \sum_{i=1}^{m} (f_A e^{-t_i/T_A} + f_B e^{-t_i/T_B} - M(t_i)/M_0)^2 \tag{4.12}$$

by choosing appropriate fitting constants f_A, f_B, T_A, and T_B. This is a nonlinear fit procedure. Sometimes, it is extended to four- or even five-exponentials, and may include also the signal at infinite time as a regression parameter. This procedure worked well for fitting most relaxation data but the interdependency of the amplitudes and time constants made it difficult to use for pore size estimation and comparisons of measurements on samples under different saturation conditions.

Another early approach, a stretched exponential fit of the following form suggested by Kenyon et al. [116, 121]:

$$M(t) = M_0 e^{(-t/T_{1\alpha})^{\alpha}}, \tag{4.13}$$

is also a nonlinear fitting procedure. However, this method has not been used widely. It does not provide an intuitive picture of its connection to the pore size distribution. It is also difficult to incorporate other petrophysical analyses such as the T_2 cutoff for irreducible water saturation and the diffusion effect. Furthermore, there are numerous relaxation data for petrophysical samples that the stretched exponential form does not fit well.

It was soon found that these nonlinear inversion processes were not adequate for efficient processing of the large amount of NMR logging data. Therefore, instead of searching for the fitting constants, T_j, we assume a set of T_j which are equally spaced on a logarithmic time scale and fit for the amplitudes, f_j, only. The inversion procedure reduces to a linear inversion problem. Although this kind of linear inversion is well known to be ill-posed, it provides a consistent definition of irreducible water saturation using a T_2 cutoff, and it is very efficient in processing large amount of NMR log data. Also because there is a physical basis for assuming a continuous pore size distribution, this linear inversion method has gained wide acceptance in the NMR logging community. We shall discuss this inversion in more detail in Chapter 6.

Figure 4.4 also shows a *relaxation time distribution* with the relative amplitude f_j corresponding to relaxation time T_j, as compared to a previously inverted three exponential result. Since T_j is proportional to a/ρ where a is the pore size (see Eq. (4.8)), this relaxation time distribution curve is frequently considered as a reflection of pore size distribution.

It should be noted that Eq. (4.9) was written based on the assumption that the water in the porous rock was in the fast diffusion regime. In fact, even in the slow diffusion regime, we can still analyze the time decaying magnetization using a sum of exponentials just like Eq. (4.9), except now the relaxation time T_j is proportional to a^2/D (instead of a/ρ) which is no longer proportional to the pore size a and becomes dependent on the diffusion coefficient of water.

Surface relaxivity

To treat the relaxation time distribution as a reflection of pore size distribution, not only do we have to be in the fast diffusion regime, but also the surface relaxivity, or the average surface density of paramagnetic impurity ions for each pore, should be nearly the same over the whole rock.

Surprisingly, numerous experiments on brine-saturated rock samples (Kenyon et al. [117, 118, 119], Straley et al. [208], Loren and Robinson [146]) have shown that indeed the relaxation time distribution curve is very similar to the pore size distribution curve obtained by other techniques such as mercury injection. Naturally, such similarity is not universal, as the NMR is a measure of the size of the pore bodies, whereas the mercury porosimetry is a measure of the size of the pore throats. However, if there is a reasonably consistent factor between the pore throat and the pore body, then the NMR relaxation time distribution can be quite similar to the pore size distribution determined by mercury porosimetry. Of course, if the average surface relaxivity ρ is not fairly constant over the whole scale, then the similarity between the two distributions can be affected.

Figure 4.5 from Straley et al. [208] shows an example of the T_2 distributions of four brine-saturated sandstones versus the corresponding differential mercury injection curves. The similarity between the two data sets is remarkable. Figure 4.6 from Morriss et al. [168] shows a different example for a particular type of rock, i.e., diatomite[1]. The match between the two

[1]A type of rock formed from the siliceous shells of diatoms which are microscopic plants living in the ocean. Diatomite's mineral composition is mainly amorphous silica (opal A). It has extremely small pore sizes.

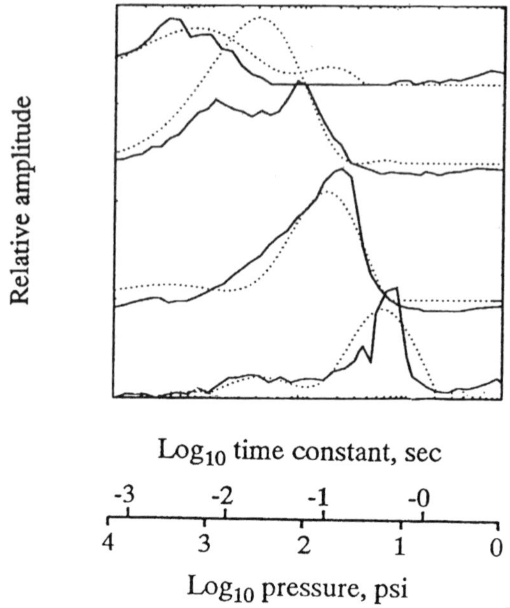

Figure 4.5: Differential mercury injection curves (solid) with the T_2 distributions (dotted) for four sandstones. (From Straley et al. [208].)

distributions is again very good until the formation of opal CT[2] which has a mercury injection pressure of 30,000 to 40,000 psi, and a corresponding T_2 near 0.4 ms, which the NMR T_2 measurements have completely missed.

By properly comparing the $T_{1,2}$ relaxation time scale

$$\frac{1}{T_{1,2}} = \rho_{1,2}\frac{S}{V} = \rho_{1,2}\frac{2}{r} \tag{4.14}$$

with the mercury entry pressure P

$$P = \frac{2\gamma\cos\theta}{r}, \tag{4.15}$$

assuming a cylindrical pore of radius r, where θ is the contact angle (140°) and γ is the surface tension (7.15×10^{-3} psi · cm) for mercury, we get

$$\rho_{1,2} = \frac{\gamma\cos\theta}{T_{1,2}P}. \tag{4.16}$$

[2]Under moderate pressures and temperatures, diatomite (opal A) becomes a partially crystallized less hydrous silica, called opal CT, before finally turning into quartz at even higher pressures and temperatures. Opal CT has even smaller pore sizes than opal A.

Thus, the surface relaxivity, based on a cylindrical model, can be easily estimated.

For Straley et al.'s example shown in Figure 4.5, the match of the two scales was found at $T_2P = 5.33$ s · psi, leading to $\rho_2 = 10.3$ μm/s. For Morriss et al.'s example in Figure 4.6, we have $T_2P = 10$ s · psi and $\rho_2 = 5.5$ μm/s.

Kleinberg [125] found that the surface relaxation strength, determined from comparing T_2 with the mercury injection curve, is very close to that determined by comparing T_2 with the air-water capillary curve. This consistency thus lends support to the validity of the general approach. He found that ρ values for many sandstones are quite consistent with the surface relaxivity found for other porous media. Other work by Borgia et al. [32] and Bryar et al. [46] have found similar ρ_1 values for sandstones. However, Bryar et al. also found some silica gel with ρ_1 as low as $1.2 \times 10^{-3}\mu$m/s. The values of surface relaxivity for various porous materials are shown in Table 4.1.

Earlier work by Howard et al. [104], using an optical method to evaluate the pore dimension, had come up with ρ values about an order of magnitude higher, apparently due to the complication of not being able to properly characterize the surface to volume ratio of the porous media. There is a more sophisticated method which extracts the surface to volume ratio from the short time behavior of the restricted diffusion coefficient $D(t)/D_0$. We shall discuss this approach in detail in Chapter 7. Using this method, Hürlimann et al. [108] found $\rho_1 = 16$ μm/s for a Fontainebleau sandstone sample.

T_1/T_2 ratio

Early laboratory NMR work on rocks was confined mostly to T_1 measurements at proton Larmor frequencies of 10 MHz and above, except for work on the NML project at Chevron in the 1950's, which covered fields corresponding to the frequency range from 8 kHz to 30 MHz. This is because the T_2 measurements are often affected by the diffusion effect caused by the internal field gradients which are difficult to characterize. In the subsequent development of pulsed NMR logging tools, it became apparent that only T_2 CPMG type measurements could offer short enough measurement time and practical logging speed. With a very short echo spacing, presumably the diffusion effect due to unknown internal field gradients in the T_2 CPMG measurements of the logging tools (Larmor frequency at 1 to 2 MHz) can be kept to a minimum.

It then becomes desirable to learn what the T_1 distribution of the rock

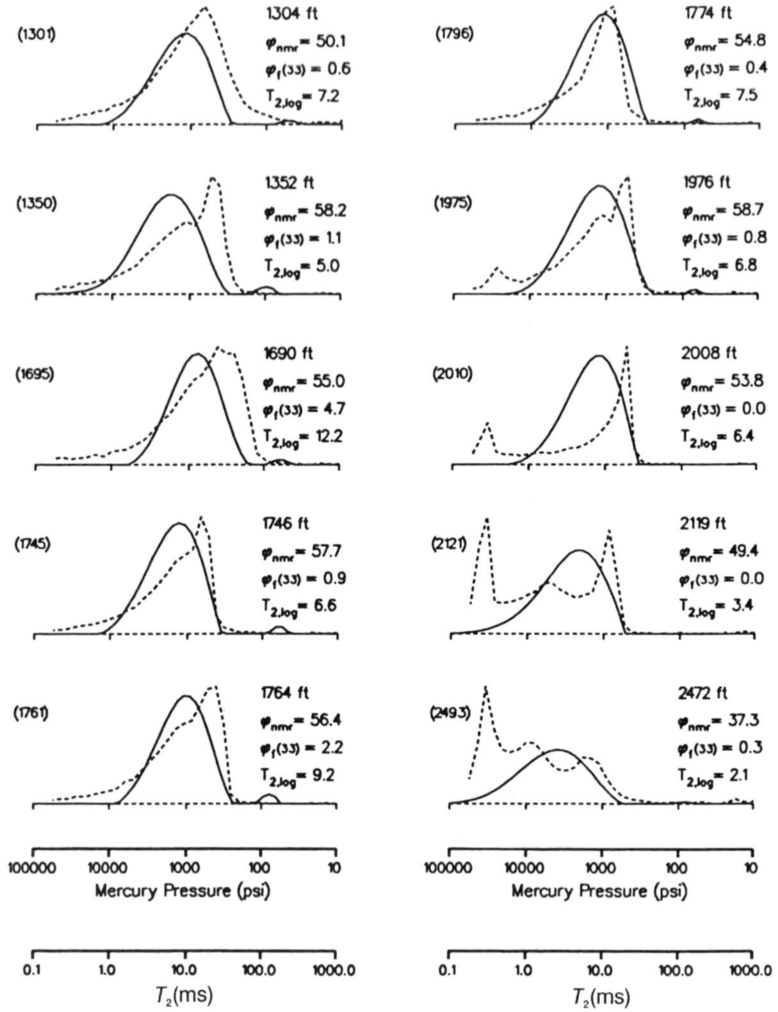

Figure 4.6: Differential capillary pressure curves (dashed) compared to T_2 distributions (solid) on cleaned and water-saturated samples from diatomite reservoir. Sample depths for the capillary pressure measurements are shown on the left in parentheses. Sample depths for the NMR measurements are shown on the right, along with NMR porosity (ϕ_{NMR}), free fluid porosity ($\phi_f(33)$), and logarithmic mean of T_2 relaxation times ($T_{2,\log}$). The opal CT, characterized by a peak near 30,000 psi mercury injection pressure on the right panel of the figure, was not detected by the NMR T_2 measurements. (From Morriss et al. [169].)

Table 4.1: Surface relaxivities of several porous materials

Material	$\rho_1(\mu m/s)$	Ref.
Quartz	3	[194]
Silica gel	1-13	[217]
Porcelain	4	[146]
Alumina	3	[35]
Glass beads	5	[35]
Glass beads	11	[207]
Ultraclean crushed SiO_2	1.5^a	[42]
Microporous porcelain	2.5^a	[33]
Sandstones	2.39^b	[32]
Sandstones	0.37^b	[97]
Sandstones	0.61^c	[195]
Silica gel	1.2×10^{-3}	[46]
Porous silica glass	1.8×10^{-3}	[72]
Silica gel	4.2×10^{-3}	[91]
Silica gel	$(3.1 - 7.7) \times 10^{-3}$	[101]
Quartz sand	0.013 ± 0.015	[46]
Quartz	0.83	[122]
Synthetic Ca silicate	$0.406 - 4.04$	[85]
Silica sand	$2.89 - 3.06$	[101]
Sandstones	9.0-46	[186]
Fontainebleau sandstone	16	[108]

[a]computed with monoexponential.
[b]computed by initial rate.
[c]computed by geometric mean of relaxation rates.

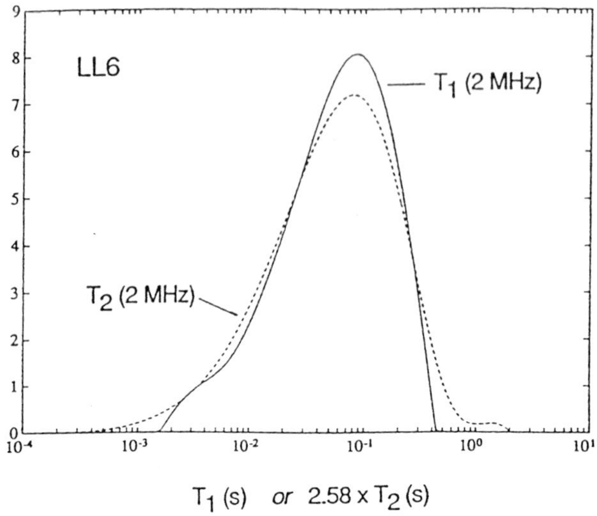

Figure 4.7: An example of T_1/T_2 cross correlation. The T_2 distribution has been shifted by a factor of 2.58 to the right to most closely match the T_1 distribution. (From Kleinberg et al. [127].)

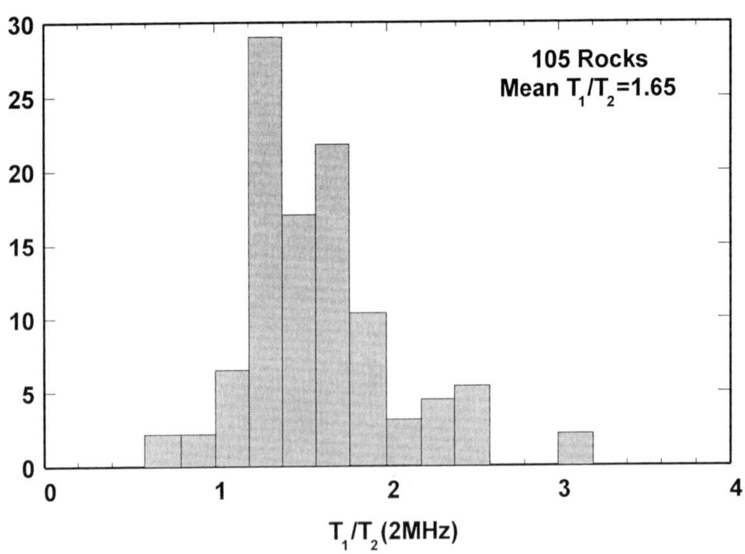

Figure 4.8: Histogram of 2 MHz T_1/T_2 ratio for 105 rocks. The CPMG echo spacing is 0.16 ms. (From Kleinberg et al. [127].)

looks like at the same Larmor frequency (1 to 2 MHz) at which we measured the T_2 distribution, what the differences between the two are, and what causes those differences.

Kleinberg et al. [126, 127] investigated many rock samples and found that when sufficiently small echo spacing is used, the diffusion effect for T_2 at 1 to 2 MHz Larmor frequency can be minimized, and that the T_2 distribution has a similar appearance to that of T_1 with a slight shift of the scale. Figure 4.7 shows such a comparison where the T_2 scale is shifted to longer times by multiplying by a factor of 2.58. Figure 4.8 shows the result of such an investigation for 105 rocks at a Larmor frequency of 2 MHz. The mean T_1/T_2 ratio is about 1.65.

The T_2 relaxation process in rocks is controlled by a number of mechanisms. Bulk relaxation can often be neglected. The surface relaxation mechanism is the same as the one that dominates the T_1 relaxation process in rocks, and provides the link to the pore size distribution of rocks. The third mechanism is molecular diffusion in macroscopic and microscopic field gradients. The CPMG pulse sequence and low Larmor frequencies can be used to minimize this diffusion effect. However, the internal gradients, induced by magnetic susceptibility contrasts between the pore fluid and solid matrix, occur at the pore scale and are difficult to eliminate completely by the CPMG pulse sequence. This kind of internal field gradient due to magnetic susceptibility contrast is proportional to the applied field strength. The enhanced relaxation rate it brings about is proportional to the square of the applied field strength. It is the primary source for the difference between T_1 and T_2 distributions [19]. As the Larmor frequency increases, the T_1/T_2 ratio is expected to increase.

Kleinberg et al.'s work [126, 127] has shown that aside from special circumstances such as the existence of extremely large internal field gradients, T_2 distribution can reflect the pore size distribution of rocks with limitations similar to those in the case of T_1 distribution. Their work also provides the important information that the T_1/T_2 ratio at 2 MHz is roughly around 1.65. This information can be used for porosity corrections in data inversion process when the repeat delay time between excitations during logging is not sufficient to have a complete polarization recovery of the spins.

4.1.3 NMR behavior of different rock types

Clastics

The T_1 or T_2 distributions of sedimentary rocks have been studied extensively. In such rocks, consistent NMR behaviors have been observed and correlated to petrophysical parameters of interest to the petroleum industry. Laboratory measurements [127] have shown that sandstones typically have a paramagnetic ion content of 1% and grain susceptibilities $\sim 10^{-5}$ cgs/cm^3. The surface relaxivity typically falls in a narrow range of 1-10 μm/s [125, 32]. The T_2 distribution is often highly correlated to the pore size distribution. When the grains are well sorted, the T_2 distribution can display a single peak at long relaxation times. Typically, a short relaxation time component is present as a shoulder or a separate bump indicating the presence of fine silt with capillary bound water or clays. The permeability formula developed from laboratory measurements usually works quite well for predicting permeability from NMR logs.

Carbonates

Carbonates usually have a surface relaxivity almost an order of magnitude smaller than that of sandstones. As a result, the diffusion coupling between pores is stronger. Thus, different pore scales, which would have different relaxation times in sandstones, are coupled to each other through the diffusion process, and merge to yield an intermediate state, or in the extreme case, a single relaxation time characteristic of the combined surface to volume ratio. Hence, the T_2 distribution of a carbonate may not reflect the pore size distribution of the rock.

Ramakrishnan et al. [183, 184], using a random walk simulation, were able to demonstrate that, indeed, the surface relaxivity plays an important role in determining whether the NMR relaxation time distribution reflects the pore size distribution. If the surface relaxivity is weak, spins diffuse easily between large and small pores, there can be a strong coupling of the decay of the magnetizations for large and small pores, and the NMR relaxation time distribution no longer reflects the pore size distribution.

They suggested a model for the analysis of carbonates which consist of a three-dimensional array of spherical grains surrounded by intergranular pore space, where all spherical grains are microporous, with a porosity of ϕ_μ. The total porosity of the system is ϕ, which is the sum of the intergranular pore space and the micropores of the grains. Assuming that the surface relaxivity ρ is the same for both the fluid/solid interfaces in the intergranular pores and the microporous pores, they were able to show that the apparent relaxivity ρ_a of the fluid/solid interface in the intergranular pores is given by the following expression

$$\rho_a \approx \rho(1-\phi_\mu) + \sqrt{\frac{\phi_\mu D_0}{F_\mu T_{2\mu}}} \approx \sqrt{\frac{\phi_\mu D_0}{F_\mu T_{2\mu}}}, \tag{4.17}$$

where D_0 is the diffusion coefficient of the pore fluid, F_μ is the local formation factor for the microporous grains, and $T_{2\mu}$ is the local T_2 relaxation time for the micropores (i.e., $1/T_{2\mu} = \rho S_\mu / V_\mu$, S_μ and V_μ being the surface and pore volume of the microporous medium). The second term is much larger than the first term, hence, the apparent relaxivity can be approximated by the second term.

With this model, the apparent relaxivity becomes temperature sensitive through the diffusion coefficient. They argued that this explains the temperature dependent behavior of the T_2 distribution for carbonates of grain-supported lithology when two pore length scales exist. For carbonates with mud supported lithology which also show strong temperature dependence in their T_2 distributions, they speculated that the intrinsic relaxivity ρ itself may be temperature dependent [86].

Thus, the temperature dependent behavior of the T_2 distributions for carbonates has not been fully understood yet. Some of them may be simply due to the fact that the temperature sensitive bulk relaxation time cannot be neglected for carbonates which have large pore sizes and a small surface relaxivity. One of the consequences of the temperature dependent behavior of the T_2 distributions is that the T_2 cutoff (see Section 4.3.2) determined at room temperature in the laboratory may need modifications in the analysis of the T_2 data acquired at the reservoir temperature.

Clay

The short relaxation time components in a T_2 distribution of a rock is frequently attributed to the water in clays. It is common practice to categorize the T_2 components below 3 ms as being associated with clay bound water, and those between 3 and 33 ms to be irreducible capillary bound water, and those above 33 ms to be free fluid. Clearly, this is an oversimplified partition. Earlier, Straley et al. [208] compared the clay-bound water volume (computed from cation exchange capacity/pore volume, Q_{ve}, which was determined from electrical membrane measurements) with the NMR T_2 measurements at 2 MHz, for 45 oilfield sandstones, and determined that the T_2 cutoff at 3 ms provided the best correlation between the two. Later, Prammer et al. [177] reported measurements of T_2 relaxation times for various types of clay, and showed that these clays cover a wide range of relaxation times. Their results (shown in Table 4.2) indicate that kaolinite has T_2 relaxation time peaks varying from 8 to 16 ms, whereas illite has T_2 peaks around 1 to 2 ms and montmorillonite has T_2 peaks from 0.3 to 1 ms.

Prammer et al.'s work [177], though pointing out that the convenient T_2 cutoff at 3 ms for categorizing clay is somewhat arbitrary, does suggest the possibility that the T_2 peak position can be used for clay typing.

Further work by Matteson et al. [156] suggested that such clay typing may be difficult. They conducted T_1 and T_2 measurements of brine mixed with four clays (kaolinite, smectite, illite, and glauconite) as function of compaction, at a Larmor frequency of 2 MHz and an echo spacing of 0.16 ms (similar to those used in Straley et al.'s experiments [208]). The measurements were started on clays in a state of slurry and proceeded to clays subjected to a compaction pressure of 16,000 psi. Their results (shown in Figure 4.9) indicate that there is only one relaxation peak, regardless of clay type, which shifts from over 10 ms continuously to below 1 ms as the compaction pressure increases. Thus, they concluded that it is not possible to do clay typing from the NMR T_2 relaxation time distributions if the compaction history is not known.

From their experiments, the shift of the T_2 relaxation time appears to be simply reflecting the change of pore volume (surface area, presumably, should stay constant) as a function of the compaction. Assuming the surface area of the clay stays the same during compaction, the computed surface relaxivity remains nearly constant as a function of compaction. Among the four clays they studied, kaolinite has the lowest value for the surface relaxivity, 1.8 μm/s, and glauconite has the highest, 3.3 μm/s. The T_2 peak positions of the slurry and centrifuged samples are, from the longest to shortest relaxation time, kaolinite, smectite, illite, and glauconite. For compacted samples from 500 to 16,000 psi, the kaolinite has the longest relaxation time, whereas smectite, illite, and glauconite have the same peak position which is shorter than that of kaolinite.

One of Matteson et al.'s findings was that there does not exist a separate peak as clay-bound water. All the observed peaks appear to be simply due

Table 4.2: Prammer et al.'s laboratory T_2 measurements at 1 MHz and at 25°C of various types of clay. Samples were saturated with different amounts of brine, pressed at 2,500 psi, and sealed. Water weight was determined from weight loss by overnight drying in a vacuum chamber at 103°C. The apparent transverse surface relaxivities ρ_2 are calculated from specific surface area, the water-to-dry weight ratios, and T_2 values. All measurements were performed at an echo spacing of 0.5 ms.

Clay Type	Water Wt[a] (%)	Signal[b] (%)	T_2[c] (ms)	ρ_2[d] (μm/s)	Remarks
Montmorillonite	7.0	20	-	-	$T_2 < 0.2$ms
	18.9	90	0.3	1	
	31.1	100	0.5	1	
	54.4	100	1	0.9	$T_1 = 1.5$ms
Illite	8.8	90	1	0.9	
	15.8	100	2	0.8	
Kaolinite	11.7	100	8	0.8	
	17.4	100	12	0.8	
	20.0	100	16	0.7	$T_1 = 30$ms
Chlorite	7.5	100	5	0.4	ρ_2 too low[e]

[a]Water weight per dry clay weight.
[b]Ratio of measured NMR amplitude per milliliter water content.
[c]T_2 of clay bound water.
[d]Apparent transverse surface relaxivity.
[e]ρ_2 may be too low as a result of overestimated surface area.

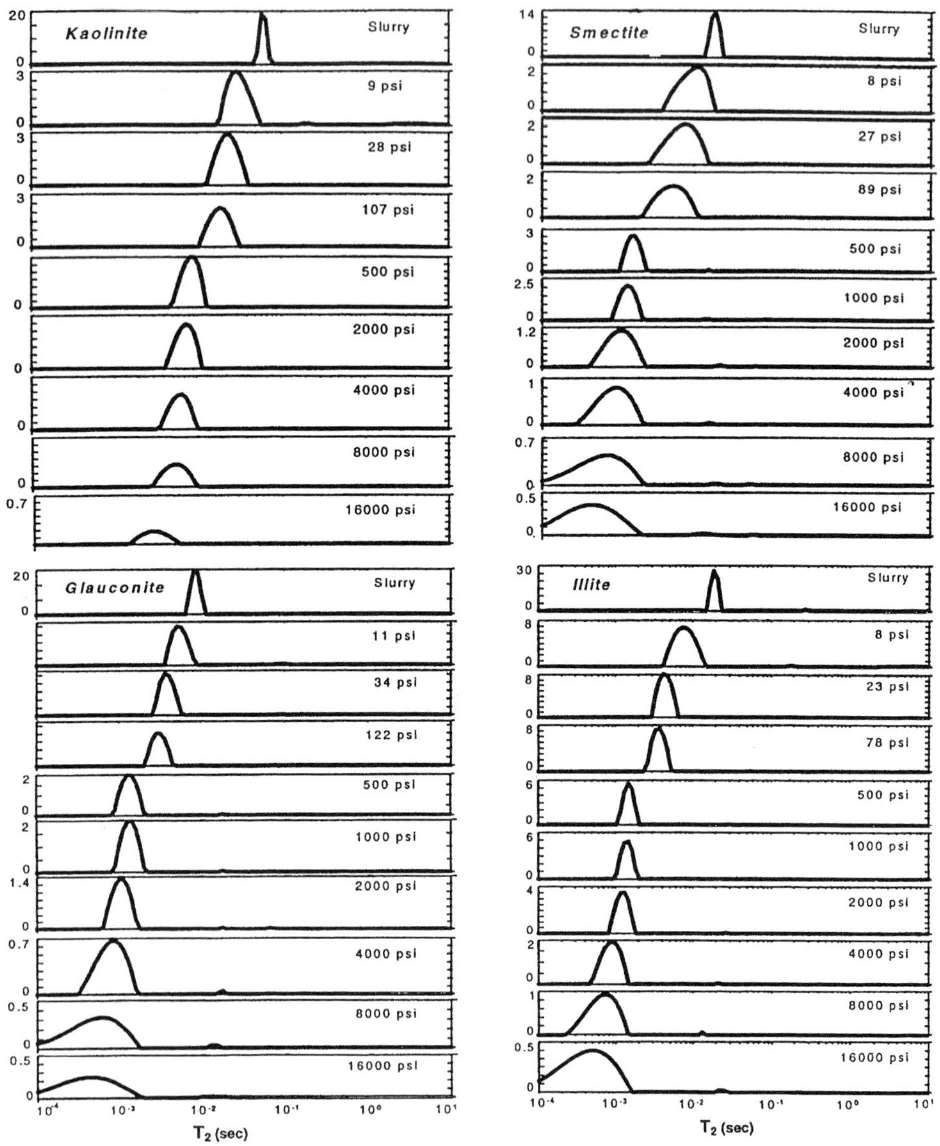

Figure 4.9: The T_1 and T_2 distributions of clay-brine slurries for kaolinite, smectite, illite, and glauconite. (From Matteson et al. [156].)

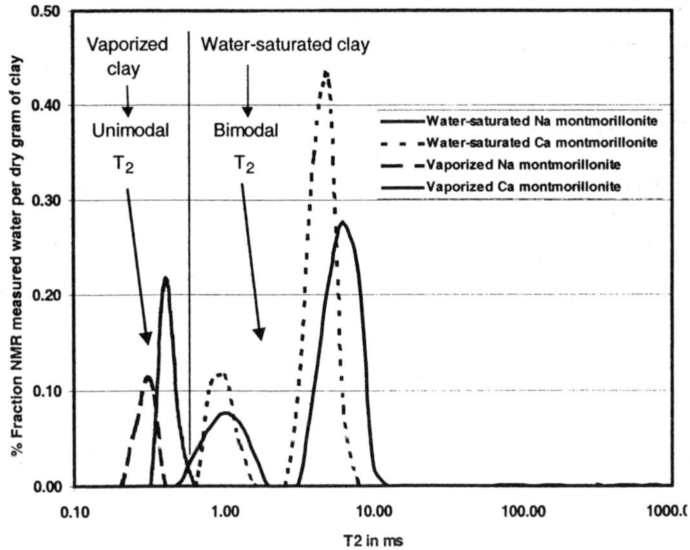

Figure 4.10: Typical T_2 distributions of hydrated montmorillonite. (From Chitale et al. [61].)

to the pore water between the clay particles. Chitale et al. [61], on the other hand, found that for hydrated samples of Ca and Na montmorillonite there exists a fast T_2 component clustered around 1 ms. As the water content increases beyond the critical saturation for the clay, the T_2 distribution becomes bi-modal, with the fast component staying the same near 1 ms and the slow component moving toward longer time. Chitale et al.'s result is shown in Figure 4.10. While there are differences between the experimental findings, it seems fair to say that kaolinite has the longest relaxation time under the same compaction history. Thus, at least distinguishing kaolinite from other clays appears possible.

4.2 Diffusion in porous media

Near the end of Chapter 2, we discussed the effect of diffusion of spins in a constant field gradient in an infinite medium. In a Hahn's echo measurement, the spin dephasing due to diffusion causes an additional magnetization decay besides the normal T_2 relaxation. The total decay is given by the following expression

$$\exp\left(-\frac{2\tau}{T_2} - \frac{2}{3}\gamma^2 G^2 D \tau^3\right), \tag{4.18}$$

where 2τ is the echo spacing, γ is the gyromagnetic ratio, G the constant field gradient, and D the self diffusion coefficient of the molecule where the spin resides. For a CPMG type measurement, the magnetization decay of the nth echo is given by the following expression

$$\exp\left(-\frac{2n\tau}{T_2} - \frac{2}{3}\gamma^2 G^2 D n \tau^3\right). \tag{4.19}$$

Diffusion only affects T_2, not the T_1 measurement[3], because there are spin dephasing and refocusing involved in the former.

We should first realize that, if the field is perfectly homogeneous everywhere, the diffusion of spins will not produce any degradation to the magnetization. This is because all spins, no matter where they go, precess at the same Larmor frequency, and there is no dephasing. If there is a field gradient and no diffusion, then the dephased spins can be brought back in phase by a π pulse for refocusing in a spin echo experiment. In reality, neither a perfectly homogeneous field nor a non-diffusive spin is possible. Thus, when both field gradient and diffusion are present, dephasing will occur and cause further degradation in the T_2 magnetization decay. Eqs. (4.18) and (4.19), which describe these decays, are strictly valid only for an infinite fluid medium, and are only approximately correct in a porous medium under conditions of restricted diffusion.

4.2.1 Enhanced T_2 relaxation

As we move from relaxation in an infinite fluid medium to relaxation in a fluid-saturated porous medium, analysis becomes more complicated. However, as long as the diffusion time is short, i.e., the τ in Eq. (4.19) is small, we can assume most of the spins do not experience the presence of the pore wall. Thus, Eq. (4.19) derived for an infinite uniform fluid is still approximately correct for the pore fluid in the rock. We can write the measured apparent T_2 relaxation rate of the rock in the following form:

$$\frac{1}{T_2} = \frac{1}{T_{2B}} + \frac{1}{T_{2S}} + \frac{1}{T_{2D}} \tag{4.20}$$

[3]See footnote [1] of Chapter 2 in p. 29

where the $1/T_2$ in Eq. (4.19) is replaced by $1/T_{2B} + 1/T_{2S}$, and the enhanced relaxation rate is replaced by $1/T_{2D}$.

T_{2B} represents the T_2 value for the bulk fluid, and T_{2S} represents the relaxation caused by the surface relaxivity which we discussed in the previous section, i.e.,

$$\frac{1}{T_{2S}} = \rho_2 \frac{S}{V}. \tag{4.21}$$

The enhanced T_2 relaxation rate, caused by the effect of spin diffusion in a field gradient, is given by:

$$\frac{1}{T_{2D}} \equiv \frac{(\gamma G \tau)^2 D}{3} = \frac{(\gamma G t_E)^2 D}{12}, \tag{4.22}$$

where $t_E = 2\tau$ (TE) is the time between echoes.

In general the value of $1/T_{2B}$ is much smaller than $1/T_{2S}$, and usually can be neglected. Then the relaxation rate $1/T_{2S}$ is enhanced by the presence of $1/T_{2D}$. For porous systems such as rocks, which have multiple pore sizes, T_{2S} is replaced by a distribution of T_2 values representing different pore sizes, and each of them is modified by a constant enhanced relaxation rate $1/T_{2D}$ when the rock is placed in an externally applied constant field gradient and there are no significant internal gradients. Thus, a large T_{2S} experiences a large shift in the apparent T_2, whereas a small T_{2S} acquires only a very small shift in its apparent value.

As we stated earlier, Eq. (4.20) is approximately valid for fluid-saturated rocks as long as the diffusion time (i.e., TE or 2τ) is very short. For the same length of diffusion time, a large field gradient causes much faster dephasing. Naturally, the length of this diffusion time, for which Eq. (4.20) is approximately valid, depends on the value of the externally applied constant field gradient. The larger the value of the field gradient, the smaller the value of the diffusion time for which Eq. (4.20) is approximately valid.

Under such constraints, Eq. (4.20) is frequently used for the analysis of T_2 distributions of rocks in a well-defined externally applied constant field gradient. Note that $1/T_{2D}$ cannot properly characterize the enhanced T_2 relaxation rate caused by the internal local field gradients within the pore space, because the latter are not uniform at the pore scale.

Thus, when the diffusion time is long, there are at least two additional considerations we need to think about in going from the infinite uniform fluid to a fluid-saturated porous medium, i.e., internal field gradients and restricted diffusion. These two issues are very complicated and difficult to treat theoretically in a rigorous manner. We shall study these in more detail in Chapter 7. For now, we will only review these qualitatively, and discuss the general approximate approaches by making several assumptions which the logging environment may or may not satisfy.

4.2.2 Restricted diffusion

As the spins diffuse in the pore space, they encounter the pore wall of the solid grains and redirect their random walk path to other directions in the pore space. Certainly, the diffusion profile will differ from the profile of spins which are in an infinite medium and are not restricted by the pore geometry. As a result, the spin echo amplitude, i.e., Eqs. (4.18) and (4.19), will be modified accordingly. Since these equations are frequently used in NMR log interpretation for estimating the diffusion coefficient of the oils or the field gradients in the rocks, it is important to learn what needs to be modified and what are the bounds of validity for these equations.

At times short enough that most spins do not encounter the pore walls or experience a significant change in local gradient, we expect Eqs. (4.18) and (4.19) to remain valid. As time goes on, since diffusion is restricted, we expect the effective diffusion coefficient $D(t)$ to decrease with time. In the short time regime, only spins within a distance $\sqrt{D_0 t}$ from the pore walls, of area S, sense the presence of the walls. Thus, only a fraction $\sqrt{D_0 t}S/V_p$ of the total spin population, where V_p is the pore volume, deviate from their diffusive behavior in the free medium. Thus the diffusion behavior is expected to deviate from its free medium behavior by the same order of magnitude. Mitra et al. [165, 166, 167], using an approximate solution for the diffusion propagator near a pore wall, were able to derive the following temporal behavior for restricted diffusion in the short time regime:

$$\frac{D(t)}{D_0} = 1 - \frac{4}{9\pi}\frac{S}{V_p}\sqrt{D_0 t} - \frac{S}{12V}\left\langle\frac{1}{R_1}+\frac{1}{R_2}\right\rangle D_0 t + \frac{1}{6}\frac{\rho S}{V_p} D_0 t + \mathcal{O}[(D_0 t)^{3/2}] \quad (4.23)$$

This relation was later used by Hürlimann et al. [108] and Slijkerman and Hofman [202] to obtain the surface relaxivity experimentally.

For a long diffusion time, the time dependent diffusion constant approaches a constant value if the pore space is connected. If the surface relaxivity ρ is zero, then the asymptotic value of the effective diffusion coefficient is related to the electrical conductivity via the standard Einstein relation [75]

$$\lim_{t\to\infty} \frac{D(t)}{D_0} = \frac{1}{F\phi} \tag{4.24}$$

where $F \equiv \sigma_p/\sigma_e$ is the formation factor[4] of the porous medium with σ_p being the conductivity of the pore fluid (brine) and σ_e being the effective conductivity of the porous medium, assuming the rock matrix is non-conductive. When $\rho \neq 0$, the infinite time limit of $D(t)/D_0$ is related to another electrical conductivity problem, where the pore fluid has a modified electrical conductivity – see Section 7.1.2, Eq. (7.35), and Ref. [21].

4.2.3 Spin echo amplitude

To derive the T_2 spin echo amplitude in a restricted geometry, one has to evaluate the phase shift accumulated by a precessing spin in a Hahn spin echo experiment:

$$\Phi = -\gamma\Big(\int_0^{\tau} dt\delta H[\mathbf{r}(t)] - \int_{\tau}^{2\tau} dt\delta H[\mathbf{r}(t)]\Big) \tag{4.25}$$

where δH is the local field deviation from a uniform background field H_0

$$H(\mathbf{r}) = H_0 + \delta H. \tag{4.26}$$

The average spin echo amplitude is evaluated by

$$\langle e^{-i\Phi}\rangle = \frac{1}{V_p}\int\int d\mathbf{r}d\mathbf{r}'G(\mathbf{r}\mathbf{r}'|t)e^{-i\Phi} \tag{4.27}$$

which takes into account all possible paths of a diffusing spin by means of the diffusion propagator $G(\mathbf{r}\mathbf{r}'|t)$ and averages over all spins in the pore space. The diffusion propagator $G(\mathbf{r}\mathbf{r}'|t)$ is the solution of the following diffusion equation:

$$\begin{aligned}
D\nabla^2 G(\mathbf{r}\mathbf{r}'|t) &= \frac{\partial G(\mathbf{r}\mathbf{r}'|t)}{\partial t} && \text{in the pore space} \\
\hat{\mathbf{n}}\cdot D\nabla G(\mathbf{r}\mathbf{r}'|t) + \rho G(\mathbf{r}\mathbf{r}'|t)\Big|_S &= 0 && \text{at the interface} \\
G(\mathbf{r}\mathbf{r}', t=0) &= \delta(\mathbf{r}-\mathbf{r}') &&
\end{aligned} \tag{4.28}$$

[4]Formation factor is a quantity frequently used in petrophysics, and it is related to another frequently used petrophysical quantity, tortuosity α, through the relation $\alpha = F\phi$, where ϕ is the porosity.

Neuman [171] carried out an approximate version of this calculation for the simple geometries of parallel plates, sphere, and cylinder in a constant field gradient, by evaluating $\langle e^{-i\Phi} \rangle$ under the assumption that the phase Φ is distributed as a Gaussian. He showed that when τ (half of the echo spacing) is much less than a^2/D, i.e., the diffusion time across a pore dimension a, the magnetization reduces to the same form as in the case of the infinite medium:

$$M(2\tau) \sim \exp\left(-\frac{2\tau}{T_2} - \frac{2}{3}\gamma^2 G^2 D \tau^3\right) \tag{4.29}$$

This is what we expected. In fact, in a restricted geometry such as porous media like rocks, we have to satisfy two requirements before we can safely use Eq. (4.29). First, the distribution of the phase (Eq. (4.25)) should be Gaussian or nearly Gaussian. The Gaussian approximation is good only for a limited range of diffusion time and field gradient strength. Beyond that, the distribution is no longer Gaussian, and an equation of the form of Eq. (4.29) is no longer valid. Second, when the Gaussian approximation is valid and a magnetization decay form of Eq. (4.29) is used, we need to replace D with a time dependent diffusion coefficient $D(t)$ when the diffusion time t is long. As this depends on many factors, such as surface relaxivity, pore size, field strength, etc., it is difficult to set a guideline. However, it is not unusual that with an echo spacing of 2 ms, the diffusion coefficient may experience a 10 to 20% reduction due to restricted diffusion.

A more rigorous discussion of this problem appears in Section 7.1.3.

4.2.4 Internal gradients

The internal field gradients of porous media are induced mainly by the magnetic susceptibility contrast between the solid matrix and pore fluid. The magnitudes of these internal gradients depend not only on the magnetic susceptibility contrast but also on the pore geometries. The local field gradient can be extremely high near the sharp corners of solid grains. When spins diffuse near those corners, they can be quickly relaxed.

Strong internal field gradients present at least two problems. First, T_2 of the short relaxation time components can relax so fast that the signal is not captured, leading to an underestimated porosity. Second, the shift of the T_2 peaks towards shorter relaxation times can cause errors in the computation of the diffusion coefficient or viscosity of the oils, or other errors in the interpretation of the T_2 distributions. Thus, it is important to understand the effect of strong internal field gradients and further to quantify such an effect.

There have been numerous attempts to quantify such internal field gradient distribution (Bendel [12], Hürlimann [109], Appel et el. [10], Zhang et al. [231]). But the complexity of the problem makes rigorous analysis of experimental data extremely difficult. We attempt to express the CPMG measurement result for a fluid-saturated rock with internal field gradients as follows:

$$\frac{M(t_i)}{M_0} = \sum_j f_j \exp\Big[-\frac{t_i}{T_j}\Big] \int dg P_j(g) \exp\Big[-\frac{1}{3}\gamma^2 g^2 D_p \tau^2 t_i\Big] \tag{4.30}$$

where $t_i = 2i\tau$, j is referring to different pore sizes, f_j is the volume fraction of pore size j, and the integration is over different possible values of the field gradient g in that type of pore, with the normalized field distribution $P_j(g)$, i.e.,

$$\int P_j(g) dg = 1. \tag{4.31}$$

Here, we assume the enhanced relaxation rate is always proportional to g^2, which may not be true, and we also assume that all pores of size j have the same distribution of internal field gradients, which, of course, may be a shaky assumption as well.

The analysis of experimental data is very difficult because it is not trivial to decouple the surface relaxation term from the field gradient contribution. However, if we make the drastic assumption that $P_j(g)$ for all pore sizes are the same, the integration in Eq. (4.30) becomes a common factor for all pore sizes and can be pulled out of the summation. Then, the ratio of the magnetization measured with an arbitrary echo spacing τ to the magnetization measured with the smallest echo spacing τ_s becomes:

$$\frac{M(t_i,\tau)}{M(t_i,\tau_s)} = \int dg P(g) \exp\Big[-\frac{1}{3}\gamma^2 g^2 D_p \tau^2 t_i\Big] \tag{4.32}$$

where we assume

$$M(t_i,\tau_s) \longrightarrow \sum_j f_j \exp\Big[-\frac{t_i}{T_j}\Big] \tag{4.33}$$

and

$$\int dg P(g) \exp\Big[-\frac{1}{3}\gamma^2 g^2 D_p \tau_s^2 t_i\Big] \longrightarrow 1 \tag{4.34}$$

for the smallest τ, i.e., τ_s. Once Eq. (4.30) is reduced to the form of Eq. (4.32), it is possible to obtain the internal field gradient distribution $P(g)$ by inverting this Gaussian transformation. Obviously, many rocks do not satisfy such a requirement, and the interpretation of the distribution $P(g)$ obtained this way would then be ambiguous.

4.3 Petrophysical applications

To understand and/or interpret the results of NMR logs from borehole measurements, it is imperative to conduct NMR laboratory measurements on brine and/or oil saturated rock samples first. Such measurements can be carried out to almost any desired accuracy using various means not readily available in a logging situation. For example, the measured NMR signals can be conveniently stacked many times to obtain a very high signal to noise ratio. One can study how the level of noise tends to affect the characteristics of the T_2 distributions. When the water and oil signals are both present in the measurements, one can soak the core plugs in D_2O to remove the water signal, or use the Magic Angle Spinning (MAS) technique to differentiate the two phases in high magnetic field. The determination of irreducible water saturation, the understanding of oil T_2 relaxation times with respect to its viscosity and diffusion coefficient, as well as the establishment of permeability estimators, are all important steps in the laboratory for the understanding, interpretation, and applications of NMR logs to formation evaluation. In the following we shall discuss them in detail.

4.3.1 Porosity

The NMR signal, when extrapolated back to zero time (in a T_1 or T_2 measurement), is proportional to the population of the hydrogen nuclei in the probed zone. When properly calibrated, the NMR measurement is an excellent way to measure porosity. Figure 4.11 (from Straley [208]) shows the excellent comparison of 2 MHz NMR porosity of T_2 measurements with buoyancy porosity for 192 sandstones. However, if the T_2 measurement is conducted at very high magnetic field, the correlation can be poor. This is because the magnetic susceptibility contrast induced internal field gradients are proportional to the applied field strength, and the enhanced relaxation rate is proportional to the square of the applied field strength. Thus, as the

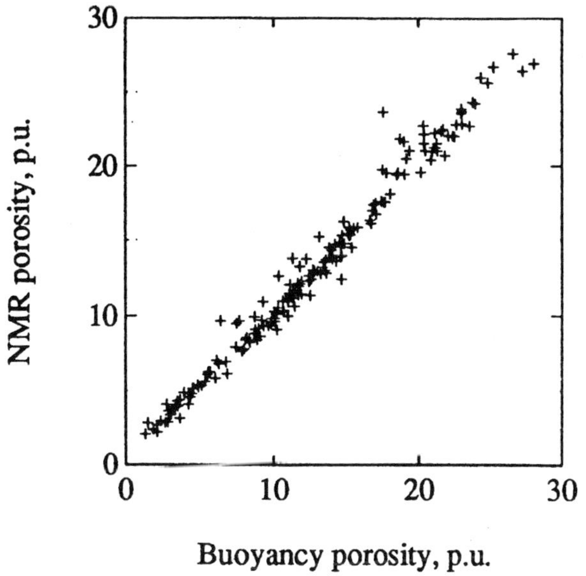

Figure 4.11: Comparison of 2 MHz NMR porosity with buoyancy porosity for 192 sandstones. (From Straley et al. [208].)

field strength increases, the increased T_2 relaxation rate causes some of the NMR signal to decay too quickly to be recorded.

Figure 4.12 shows a comparison of NMR porosity at three field strengths to buoyancy porosity. It is clear that those measured at 85 MHz suffered significant signal loss from induced internal gradients. Typically, one should do these experiments near the logging frequencies, i.e., 1 to 2 MHz. Then, with a reasonably small echo spacing, internal gradient effects can be minimized.

However, even at Larmor frequencies of 1 to 2 MHz, porosity can be underestimated when the Q of the rf coil is degraded by highly saline brine in the rock pores. This problem can be dealt with by proper calibration and adjustments to the Q of the rf coil.

4.3.2 Irreducible water saturation

One of the more important applications of NMR logging is to estimate the irreducible water saturation. Irreducible water saturation is not a precisely defined term. It can be loosely defined as the water saturation below which increases in capillary pressure will result in negligible decreases in its value.

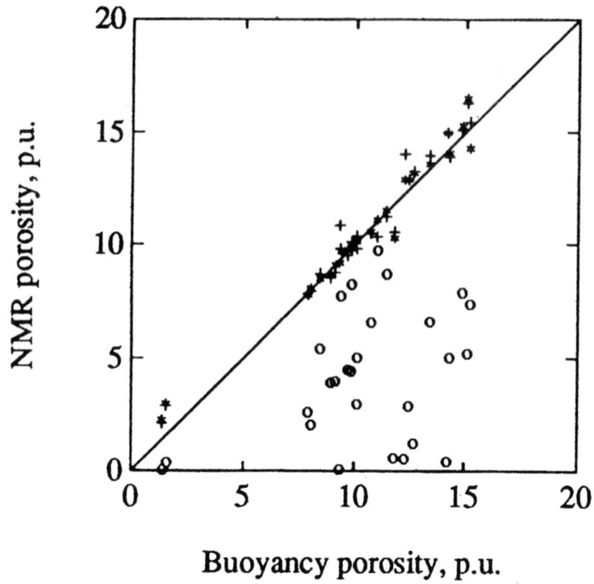

Figure 4.12: Comparison of NMR porosity of T_2 measurements at three field strengths to buoyancy porosity. o is 85 MHz, + is 2 MHz, and $*$ is 1 MHz data. (From Straley et al. [208].)

Capillary pressure is derived from the difference in density between oil and brine which causes the oil to displace or "drain" the water from the formation. In the oil industry, a capillary pressure of 50 psi is often used to determine irreducible water saturation in poorly consolidated core. For well consolidated core, a capillary pressure of 100 psi or higher is typically used.

Early experiments by Timur [213] use a three-exponential model to analyze the T_1 relaxation data for 100% fully brine saturated core samples. The fraction associated with the long T_1 component(s) (defined as Free Fluid Index, FFI) beyond a critical T_1 value was plotted against the movable fluid determined by the core measurements with a desaturation pressure of 50 psi. A good linear correlation was found. Usually, the irreducible water saturation is found to be related to the shortest T_1 component. Thus, the NMR measurements of a brine-saturated rock provide a good estimate on its irreducible water saturation.

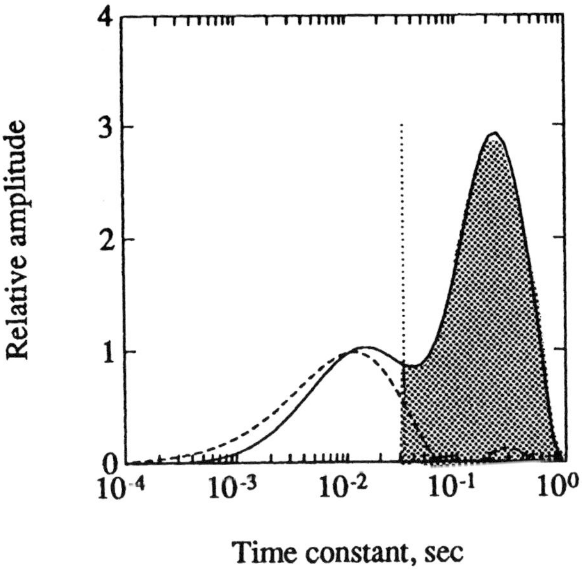

Figure 4.13: The T_2 distributions of a sandstone when it is fully saturated (solid) and when it is desaturated after centrifuging at 100 psi (dashed). The vertical cutoff in the T_2 distribution separates the porosity into the one for free fluid (shaded) and the one which is irreducible, with the latter equal to the water content of the desaturated sample. (From Straley et al. [208].)

T_2 cutoff

When a set of relaxation times equally spaced on a logarithmic scale is chosen, and a regularized inversion process is adopted for the analysis of NMR data, numerous laboratory experiments [168, 208, 76, 56, 120] indicate that a T_2 cutoff value of 33 ms for sandstones and 100 ms for carbonates are appropriate for irreducible water saturation determination. Of course, these are based on averages determined empirically from experiments. Caution should be used in the application of such cutoffs. Measurements on cores from the reservoir studied provide the most reliable cutoffs.

Figure 4.13 shows T_2 distributions of a sandstone sample when it is fully saturated and after it is desaturated by centrifuging at an equivalent air/brine capillary pressure of 100 psi. The T_2 cutoff (vertical line) separates the distribution into the porosity containing the movable fluid with long T_2 and the porosity containing the irreducible water with short T_2. Figure 4.14 shows a cross plot of the Free Fluid Index (FFI), obtained by centrifuging a suite of

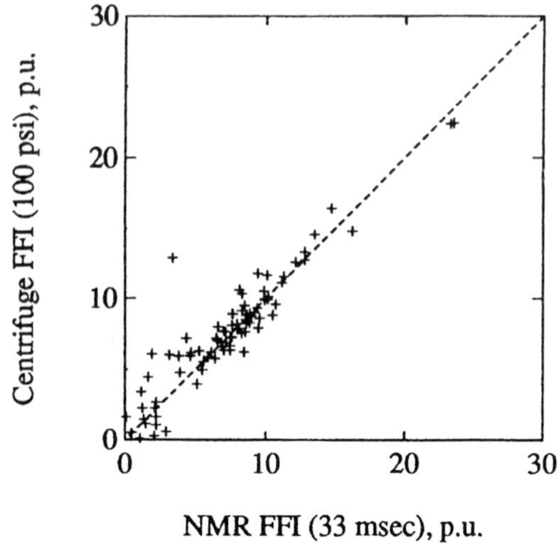

Figure 4.14: The crossplot for a suite of sandstones of centrifuge FFI at 100 psi versus the NMR determined FFI based on a T_2 cutoff at 33 ms. (From Straley et al. [208].)

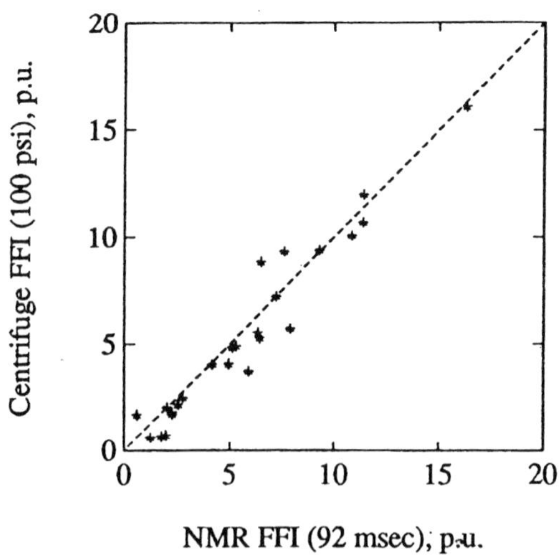

Figure 4.15: The crossplot for a suite of carbonates of centrifuge FFI at 100 psi versus the NMR determined FFI based on a T_2 cutoff at 92 ms. (From Straley et al. [208].)

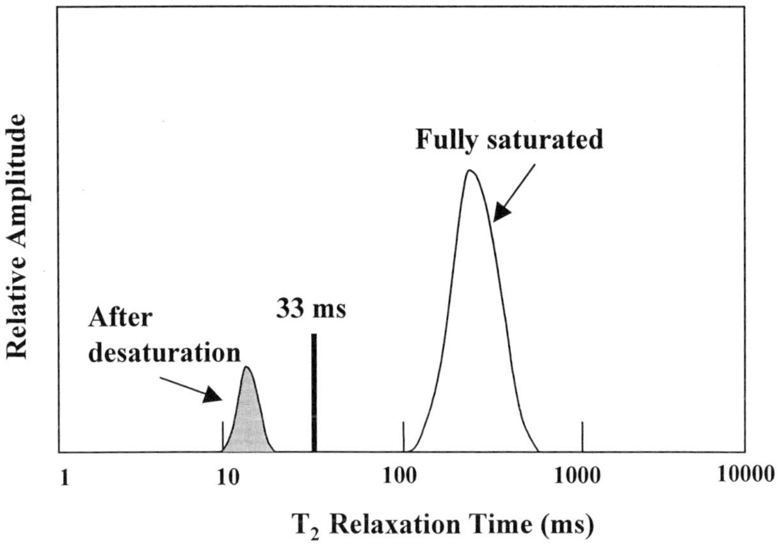

Figure 4.16: The T_2 distributions for a rock sample with a relatively uniform pore size when it is fully saturated and when it is desaturated (shaded). The T_2 peak emerges below 33 ms only after the desaturation. Note that the normal T_2 cutoff of 33 ms will not work in this case.

sandstone samples at an equivalent capillary pressure of 100 psi, versus the FFI determined by the cutoff at 33 ms on the T_2 measurements made on the same samples. Figure 4.15 shows a similar cross plot for carbonates where a T_2 cutoff of 92 ms was used. The appropriate T_2 cutoff for sandstones or carbonates can vary from reservoir to reservoir. For example, the T_2 cutoff for sandstones can vary from 20 to 50 ms, whereas that for carbonates from 90 to 150 ms. The fact that the T_2 cutoffs for carbonates are typically much larger than those for sandstones is understandably due to the much smaller surface relaxivity for carbonates.

Spectral representation

Frequently, when the pore size is uniform with little variation, the T_2 distribution for a fully water-saturated core has only one peak at long relaxation times, far above the T_2 cutoff. As the core is desaturated, the water in the pore space is emptied, leaving only small amount of water in the interstitial space near the grain contacts. This results in the emergence of a T_2 peak at a short relaxation time which was absent before desaturation. Figure 4.16 shows T_2 distributions for such a situation. In this case, the regular cutoff will not work, and it will lead to a zero value for the irreducible water.

There have been various proposals suggested to handle a situation like this. Some use a variable cutoff for different lithologies or facies, where each cutoff is determined empirically from laboratory core measurements.

Some use what is called the Spectral BVI method (Coates et al. [64]), namely, a combination of fractions of the amplitude for all the relaxation times, to determine the irreducible water saturation. For short relaxation times, nearly all amplitudes contribute to the irreducible water. For long relaxation times, only a small portion of the amplitudes contribute to the irreducible water. These different fractions of amplitude for different relaxation times are also empirically determined from core measurements.

Bi-exponential model

Another method is to use an experimentally determined empirical relation between the irreducible water saturation, S_{wirr} and the $T_{2,LM}$ (the logarithmic mean of T_2 relaxation times) for the prediction of S_{wirr}. All these methods work equally well.

Figure 4.17 is a crossplot of S_{wirr} vs $T_{2,LM}$. The data are from the NMR Rock Catalogue from the combined companies ART/SCINTEF/RESLAB in England and Norway. The S_{wirr} were obtained by centrifugation using an equivalent air/brine capillary pressure of 100 psi. The $T_{2,LM}$ were determined at an echo spacing of 0.35 ms using a 2 MHz, homogeneous field spectrometer. The data is fitted using a bi-exponential model. The fitted equation is given by:

$$S_{wirr} = 0.6101 \exp(-T_{2,LM}/15.9) + 0.3688 \exp(-T_{2,LM}/276.8) \qquad (4.35)$$

where the $T_{2,LM}$ is in ms. Long $T_{2,LM}$ values are clearly associated with small S_{wirr} values but the data density at long $T_{2,LM}$ values is sparse.

The core used in the study comes from different oil companies but is largely from North Sea reservoirs. For logging tools with a field gradient, modifications to Eq. (4.35) are needed to account for diffusion effects in the field gradient.

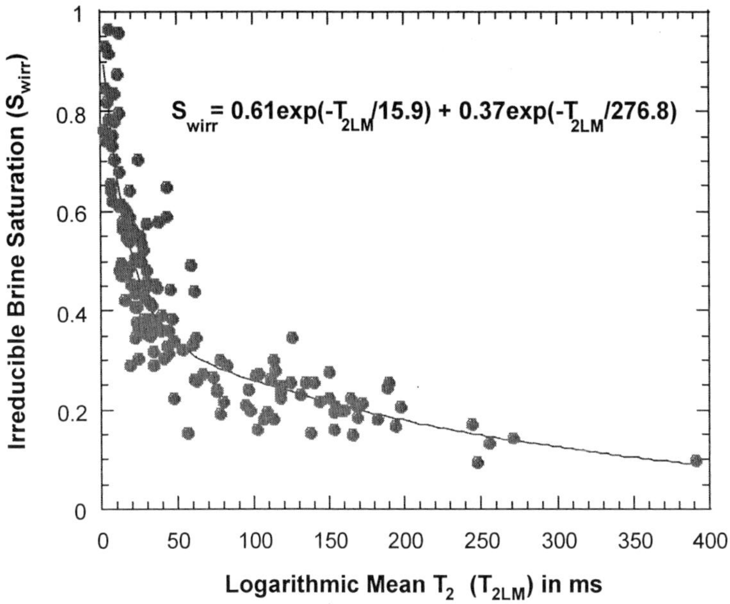

Figure 4.17: The irreducible water saturation, S_{wirr}, versus $T_{2,LM}$ of fully brine-saturated state for a suite of rock samples. The data are fitted with a two-exponential model.

Capillary curve

As we discussed earlier, the pore size distribution determined by mercury porosimetry is similar to the T_2 distribution of the rock. Similarly, its cumulative quantity, i.e., the capillary pressure curve, is connected to the cumulative T_2 distribution. This is shown in Figure 4.18. In capillary pressure based estimates of S_{wirr}, we typically use the apparent asymptotic value of S_w, beyond which very large P_c values would be needed to reduce S_w, or we determine a maximum capillary pressure for a reservoir and argue that the S_w at that pressure, in an air/brine or oil/brine drainage experiment, is the irreducible brine saturation. Such data can be interpreted in terms of a minimum pore size through which the non-wetting phase (oil) can displace the wetting phase (brine). In a similar manner, a cutoff in the T_2 distribution is often used to estimate S_{wirr}.

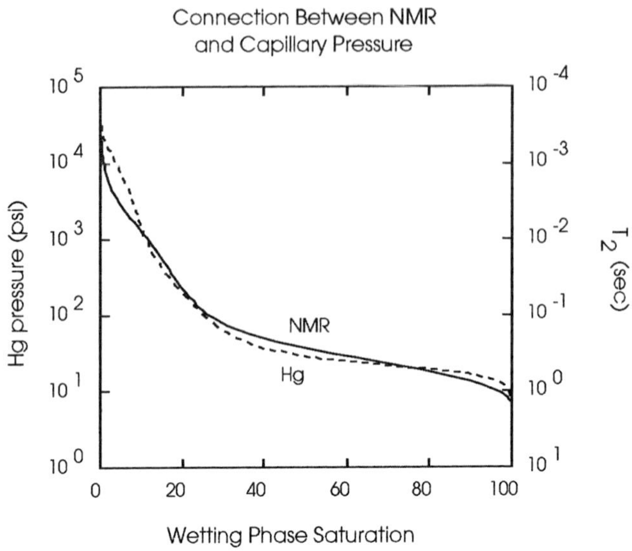

Figure 4.18: The connection between capillary pressure curve and cumulative T_2 distribution. (From Kleinberg [125].)

4.3.3 Permeability predictor

$T_{2,LM}$ equation

The fact that NMR log can be used for permeability prediction is because the T_2 relaxation time is proportional to pore size (or pore volume to surface ratio):

$$T_2 \approx \frac{V}{\rho_2 S} \tag{4.36}$$

and provides the length scale for permeability prediction. Kenyon et al. [121] and Banavar and Schwartz [11] have shown that the following equation provides a good permeability prediction for sandstones:

$$k = T_{1,LM}^2 \phi^4 \tag{4.37}$$

A general form of $k = aT_{2,LM}^b \phi^c$ can be used with the prefactor, a (which includes the $T_{1,LM}/T_{2,LM}$ ratio), and the exponents, b and c, specifically determined from fitting a set of core permeability measurements.

Other forms, such as $k = aT_{2,LM}^b / F^c$, (where F is the formation factor) have been suggested [222, 115, 113, 78, 195, 28]. Both forms work about

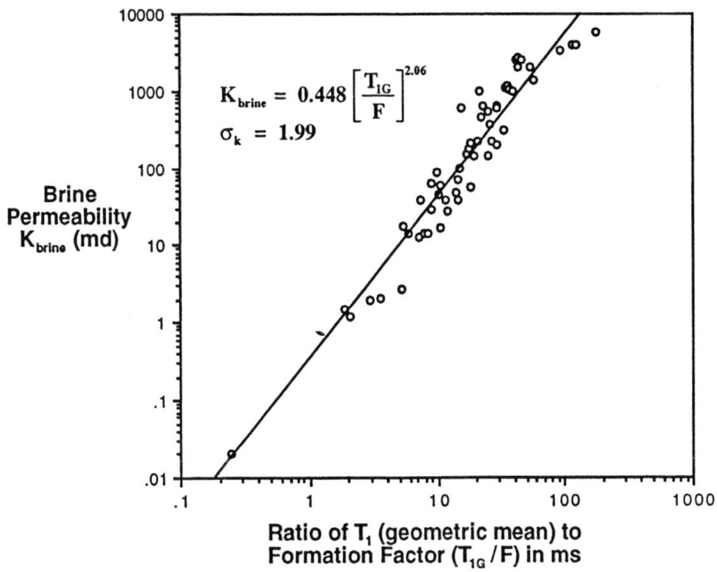

Figure 4.19: Brine permeability versus $T_{1,LM}/F$ for a suite of sandstones. The dimensionless quantity σ_k is defined in Table 4.3. (From LaTorraca et al. [132].)

equally well within experimental error. Table 4.3 [132] shows some of the comparison between different models. Figure 4.19 shows such a plot for sandstones from a single field.

This suggested form of permeability predictor has its physical basis. It was shown by Katz and Thompson [115], Johnson et al. [113], and many others, that the permeability of a porous medium can be well described by tho following form:

$$k \sim L^2(\sigma/\sigma_f) = L^2/F = L^2\phi^m. \tag{4.38}$$

Here σ and σ_f are the electrical conductivities of the fully saturated rock and the pore fluid, respectively, L is a length scale characteristic of the pore geometry, and F is the formation factor. The last part of Eq. (4.38) is simply from Archie's law (i.e., $F \sim \phi^{-m}$). As pointed out earlier, T_2 is proportional to the characteristic length of the pore scale, and hence L. Naturally, different pore geometries will affect the value of the exponents for L and ϕ. They are usually empirically determined for a suite of rocks from a reservoir.

Table 4.3: Least-squares fit permeability estimators.

Reference	Permeability (mD)	σ_k [a]	n [b]
Kenyon et al. [121]	$K_{brine} = 1.17(T_{1\alpha}^2\phi^4)^{1.13}$	2.7	
Sen et al. [195]	$K_{brine} = 0.794(T_{1\alpha}/F)^{2.15}$	3.3	
Borgia et al. [28]	$K_{air} = 3.25(T_{1\alpha}^2/F^2)^{0.79}$	2.1	
	$K_{air} = 0.076(T_{1\alpha}/F)^{0.94}$	2.3	
	$K_{air} = 10.7(T_{1\alpha}^2\phi^4)^{0.72}$	1.9	
LaTorraca et al. [132]	$K_{brine} = 0.0063(T_{1G}/F)^{1.23}$	3.7	257
	$K_{brine} = 0.43(T_{1,LM}/F)^{2.16}$	3.4	257
	$K_{brine} = 0.33(T_{1\alpha}/F)^{2.05}$	3.4	257
	$K_{brine} = 1.21(T_{1,LM}^2\phi^4)^{1.00}$	3.6	257
	$K_{brine} = 4.00(\sum_{i=1}^3 (A_i\phi)^4 T_{1i}^2)^{0.933c}$	3.7	257
(Clastic subset)	$K_{brine} = 0.29(T_{1,LM}/F)^{2.16}$	2.0	57
	$K_{brine} = 0.45(T_{1,LM}/F^*)^{2.06d}$	2.0	57

[*] K is in mD, T_1 in ms, and ϕ in fraction.

[a] σ_k is the factor by which the estimated permeability misses the measured permeability, i.e., $\sigma_k = 10^{\epsilon}$, where ϵ is obtained from $\log_{10} K = a + b\log_{10}$ (predictor) $\pm\ \epsilon$.

[b] Number of samples.

[c] A_i and T_{1i} are fitting constants in a three-exponential model fit for the T_1 distribution. Thus, $A_i\phi$ and T_{1i} are the partial porosity and the corresponding T_1 component, respectively.

[d] F^* is the clay-corrected formation factor.

Since Eq. (4.37) is determined from sandstone core plugs fully saturated with water, the presence of oil will affect the value of $T_{2,LM}$, and hence the permeability prediction. In this situation, other permeability predictors, as discussed below, using irreducible water saturation S_{wirr} as input are frequently used.

Coates/Timur equation

Multiple studies have shown that the permeability in clastic reservoirs is related to the irreducible brine saturation (S_{wirr}). Related quantities that can be estimated from NMR are the Bulk Volume Irreducible water BVI (or ϕS_{wirr}), and the Free Fluid Index (FFI) (or $\phi(1 - S_{wirr})$).

Timur [214] developed a permeability estimator based on numerous core measurements. It has the following form:

$$k = 10^4 \frac{\phi^{4.5}}{S_{wirr}^2} \tag{4.39}$$

where ϕ and S_{wirr} are in fractions, and k is in mD (milli-Darcy).

Coates [63] developed a widely used equation based on the irreducible brine saturation. There are several versions of his equation. The most commonly used one is:

$$k = ((\phi/10)^2(FFI/BVI))^2 \tag{4.40}$$

where ϕ, FFI, and BVI, are in p.u. (porosity unit), and k is in mD.

This kind of empirical form for permeability prediction can be traced back to the Kozeny equation [53], i.e.,

$$k = A\frac{\phi^3}{S^2}, \tag{4.41}$$

where ϕ is porosity, S is the surface area per unit bulk volume, and A is an empirical constant. The surface area S in the denominator is related to the irreducible water saturation or BVI.

Summation of echoes

The summation of echo amplitudes is mathematically related to the mean relaxation time, $\langle T_2 \rangle$. Sezginer et al. [199] showed that

$$p = \sum_{i=1}^{N} g_i \approx \sum_{i=1}^{N} \epsilon_i + \mathrm{HI}\phi\langle T_2 \rangle / t_E \tag{4.42}$$

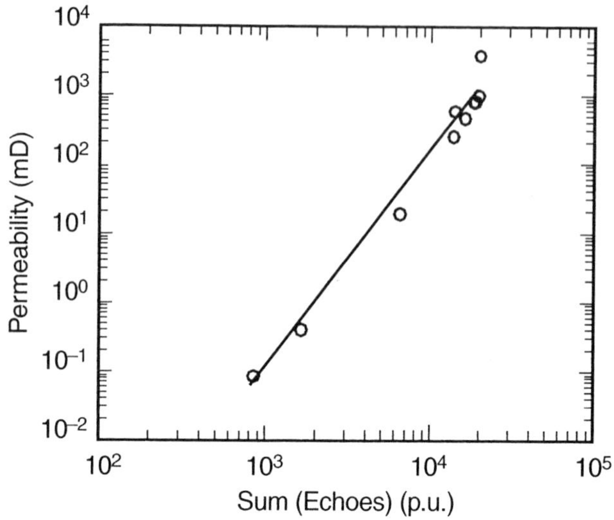

Figure 4.20: Measured permeability versus sum of echoes for 10 quarried sandstone samples. TE=0.2 ms. (From Sezginer et al. [199].)

where p is the summation of echoes; g_i, the amplitude of the i-th echo; ϵ_i, the noise level of i-th echo; HI, the hydrogen index; ϕ, the porosity; $\langle T_2 \rangle$, the mean relaxation time; t_E, the echo spacing, and N the number of echoes.

$\langle T_2 \rangle$ is somewhat different from the commonly used logarithmic mean, $T_{2,LM}$. It is related to the surface to volume ratio (with strong weight for large pores) of the rock formation; thus, it is related to permeability. From laboratory experiments on quarried sandstones, Sezginer et al. [199] found a good correlation (Figure 4.20), and suggested the following relations for TE=0.2 ms:

$$k \approx 6.18 \times 10^{-12} p^{3.52} \quad \text{for} \quad N = 600 \tag{4.43}$$

$$k \approx 4.58 \times 10^{-12} p^{3.13} \quad \text{for} \quad N = 1200 \tag{4.44}$$

where k is the permeability in mD, and p is in p.u.

Permeability prediction for carbonates

Carbonates not only occur in a wide variety of pore sizes and shapes, but also exhibit a heterogeneous distribution of pore types at all core and log-

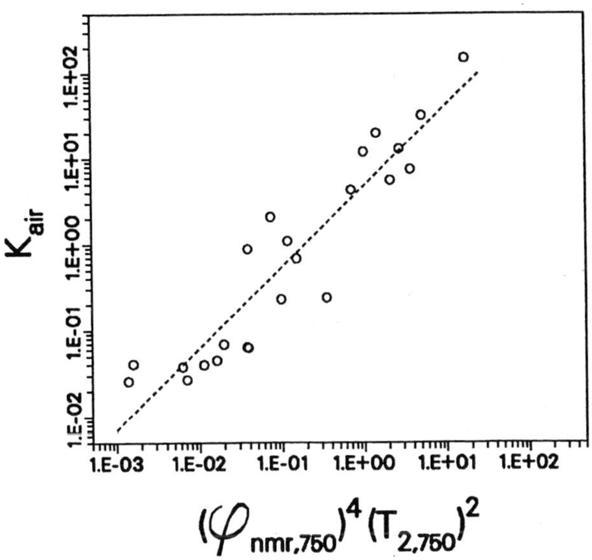

Figure 4.21: Crossplot of air permeability K_{air} versus $(\phi_{NMR,<750})^4$ $(T_{2,<750})^2$ for a suite of carbonates where the porosity and T_2 above 750 ms are not included. (From Chang et al. [56].)

ging measurement scales. Chang et al. [56] has published studies of NMR T_2 measurements on samples 100% brine saturated and desaturated by centrifugation to approximately S_{wirr}. Based on these studies, they infer that all signal above 750 ms is due to vugs, and that if the vuggy porosity is not included for ϕ and T_2, then the permeability equation $k = \phi^4 T_{2,LM}^2$ works quite well. A crossplot of air permeability K_{air} versus $(\phi_{NMR,<750})^4(T_{2,<750})^2$, for a suite of carbonates from their work, is shown in Figure 4.21. Their measurements were made in homogeneous magnetic field. The NMR logging tools usually have significant field gradients which will reduce the 750 ms cutoff substantially when the vugs are brine filled and to a lesser extent if the vugs are filled with light crude.

4.3.4 Wettability

Wettability is defined as "the tendency of one fluid to spread on or adhere to a solid surface in the presence of other immiscible fluids" [69]. The spreading on or adherence to a solid surface is commonly described as wetting. It can

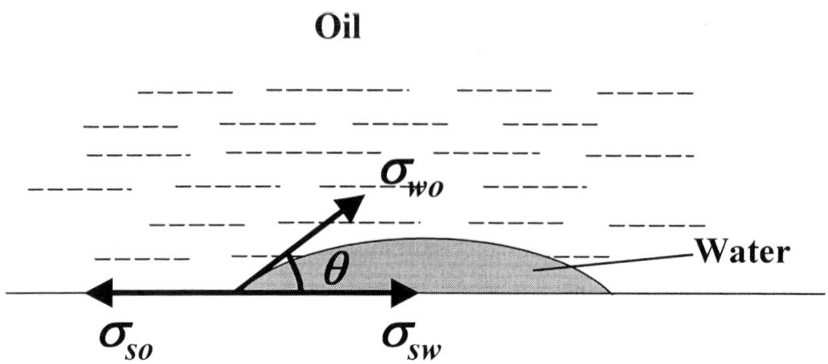

Figure 4.22: Schematic diagram of a water-oil-solid interface showing the equilibrium of interfacial forces, where σ_{so}, σ_{sw}, and σ_{wo} are the interfacial tensions between the solid and oil, solid and water, and water and oil, respectively.

be measured by *adhesion tension*, a function of the interfacial tension which determines which fluid will preferentially wet the solid.

As shown in Figure 4.22, two immiscible liquids, oil and water, are in contact with a solid. The contact angle θ is measured through the denser liquid phase and ranges from 0° to 180°. The adhesion tension A_T is defined as

$$A_T = \sigma_{so} - \sigma_{sw} = \sigma_{wo} \cos \theta_{wo} \tag{4.45}$$

where σ_{so} is the interfacial tension between the solid and lighter fluid, oil; σ_{sw} is the interfacial tension between the solid and denser fluid, water; and σ_{wo} is the interfacial tension between fluids, water and oil.

A positive value of adhesion tension indicates that the denser liquid preferentially wets the solid surface, whereas a negative value indicates the lighter liquid preferentially wets the surface. A zero value indicates both liquids have an equal affinity for the surface. A reservoir rock can be water-wet, or oil-wet, or having both conditions coexist, i.e., a mixed wettability situation. Wettability of reservoir rocks can have dramatic effects on the effective permeability of oil[5] and the efficiency of oil recovery.

There is abundant literature on wettability studies [9, 190, 170, 130]. There are different methods for measuring wettability [9], some of them quite elaborate and time consuming. Although these methods show general agreement in discriminating oil-wet versus water-wet media, there is no quantitative correspondence between these measures.

NMR relaxation is a convenient method for characterizing wettability of a pore fluid phase. When a non-wetting phase becomes partially-wetting, its relaxation time typically becomes shorter as compared to its value for the bulk phase due to surface relaxation. There have been numerous studies on the wettability problem using NMR methods [37, 191, 224, 107, 74, 29, 105, 106, 174, 27, 155, 204, 232]. There is, however, no generally agreed upon method for quantifying the wettability. These studies are largely confined to laboratory core measurements. Applications of NMR logs for wettability analysis have been scanty, and require careful coordination of core measurements and log interpretation.

Brown and Fatt [37] studied the wettability problem of reservoir rocks using an NMR method with mixtures of preferentially oil wet and preferentially water wet sands. Their result for relaxation rates of fully water saturated sand mixtures containing various percentages of preferentially oil wet sands is shown in Figure 4.23. The linear relationship clearly suggests that the water wet surface has a higher surface relaxivity than the oil wet surface, and the linear variation of the relaxation rate simply reflects the abundance of the oil wet surfaces. To apply such a finding to core analysis, they suggested using a procedure of measuring the relaxation rates of core at an unaltered wettability state, 100 percent water-wet state, and 100 percent oil-wet state. Then the percentage of the oil-wet surface for the initial mixed wettability state is determined from the straight line drawn between the 100 percent water-wet and 100 percent oil-wet states.

However, the application of such reasoning to the amount of wettability from NMR relaxation rate measurements is complicated by the pore size effect. Except when there is only one pore size, the relaxation rate measured at t=0 is an average relaxation rate of all pore sizes. If the oil-wetting occurs preferentially for larger pores, the variation of the relaxation rate may not be linear as that shown in Figure 4.23. Then the interpretation of the fractional

[5] The effective permeability is a relative measure of the conductance of the porous medium for one fluid phase when the medium is saturated with more than one fluid.

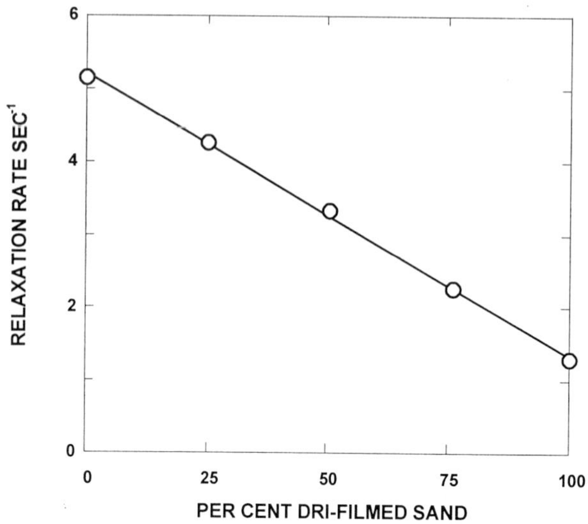

Figure 4.23: Relaxation rates for sand mixtures containing various percentages of oil wet sand. (From Brown and Fatt [37].)

wettability of a core becomes complicated and can be ambiguous.

Borgia et al. [29] and Howard [105, 106] chose chalk samples for the wettability study because of the narrow distribution of pore sizes for chalk samples making the interpretation of experimental results easier. Amott index [7] (the ratio of the spontaneous imbibition of water to the amount of imbibition plus forced displacement by centrifuge) was used in their analysis either to characterize the wettability of the rock samples or to correlate to an NMR-based wettability index. Howard suggested an NMR based wettability index as follows:

$$T_n = CS_w^b \tag{4.46}$$

where T_n is the normalized relaxation time of the water-filled component compared to a fully saturated state (i.e., $T_n = T_{2,S_w}/T_{2,S_w=1}$ where T_{2,S_w} and $T_{2,S_w=1}$ are the T_2 peak of water at partially and fully water-saturated states, respectively), C is an empirical constant, and the exponent b is correlated with wettability. For a fully water wet core, b is equal to 1. A decrease in b indicates weaker proton interaction with the pore wall, i.e., less water wet, at a given water saturation.

Zhang et al.[232] studied the wettability of some sandstones with an NMR method and found that some of them became mixed-wet after being saturated with refined or crude oil and aging for a certain length of time. Some of the pore-lining chlorite flakes apparently became oil wet and shielded the water from having contact with the pore walls.

Wettability is an important parameter in oil recovery. NMR offers a convenient technique for characterizing wettability. However, a quantitative and generally accepted method for measuring wettability has yet to be developed.

4.4 MAS NMR

The Magic Angle Spinning (MAS) technique is used routinely in solid state NMR to average out chemical shift anisotropy and internuclear dipolar interactions (see, for example, Mehring [160]). However, the application of such a technique to the study of fluids contained in rocks is relatively recent (Xiao et al. [228], de Swiet et al. [71], Wilson and LaTorraca [225]). There are many areas yet to be explored, such as oil viscosity, preferential wettability, etc., which may develop into a powerful technique for petrophysical analysis.

In fluid-saturated rocks, the paramagnetic impurities residing at the pore wall, as well as the magnetic susceptibility contrast between the solid matrix and the pore fluid, create a rather heterogeneous local field environment. Such field inhomogeneity leads to a broad featureless line in ^{1}H NMR spectra and renders it useless in identifying or quantifying the oil and water contents in the rocks. The local field $\mathbf{B}_{\text{loc}}$, in a dipole approximation, is given by

$$\mathbf{B}_{\text{loc}}(\mathbf{r}) \cdot \hat{z} = \frac{B_0}{4\pi} \int \chi(\mathbf{r}') \frac{(3\cos^2\theta - 1)}{|\mathbf{r} - \mathbf{r}'|^3} d\mathbf{r}' \tag{4.47}$$

where $\chi(\mathbf{r}')$ is the local magnetic susceptibility in the rock, B_0 the external applied field in the z direction, and θ is the angle that $\mathbf{r} - \mathbf{r}'$ makes with the applied field as shown in Figure 4.24.

When a fluid-saturated rock is placed in a rotor inclined at an angle θ with respect to the applied field B_0 in the z direction, the proton spin inside the molecules is aligned in the z direction. As the sample starts to rotate about the rotor axis, the proton spins remain aligned in the z-direction, whereas the surrounding paramagnetic atoms residing either on the pore wall or solid matrix start to rotate. Thus the proton spins experience a time varying local field which is rotating about the rotor axis. The NMR signals that

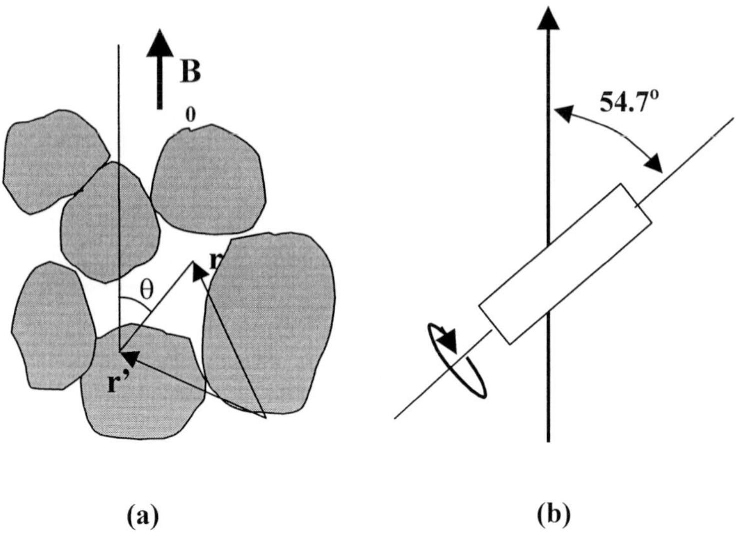

Figure 4.24: The local dipolar field in a rock (a), and the magic angle spinning (b) where the rock sample is placed inside a rotor spinning about its axis which makes an angle of 54.7° with respect to the applied field B_0.

we measure will be a time average of these varying fields. If the magnetic susceptibility $\chi(\mathbf{r}')$ is isotropic, all of these will average to zero, except the contributions of paramagnetic atoms lying along the rotor axis. But those contributions are also made to vanish, as we purposely set the inclined angle $\theta = 54.7° = \cos^{-1}(1/\sqrt{3})$ which nullifies the dipolar field given by Eq. (4.47). Hence, this technique is called magic angle spinning (MAS).

As shown in the left side of Figure 4.25, when the rotor is stationary, the ^{1}H NMR spectrum of a water-saturated Berea sandstone shows a broad featureless peak. When the rotor spins at 1 and 2.5 kHz, the broad spectrum narrows to a sharp peak with many spinning sidebands, spaced at the rotor frequency. The existence of the spinning sidebands can be understood as follows: Recall that the spectrum is the Fourier transform of the FID after the application of a 90° rf pulse. As the spins are precessing in the xy plane and signals are being collected, those with the Larmor frequency determined by B_0 will be stationary in a coordinate frame rotating about the z-axis with that frequency. All spins with different local Larmor frequencies will be rotation-averaged except those whose local Larmor frequency is higher or lower than

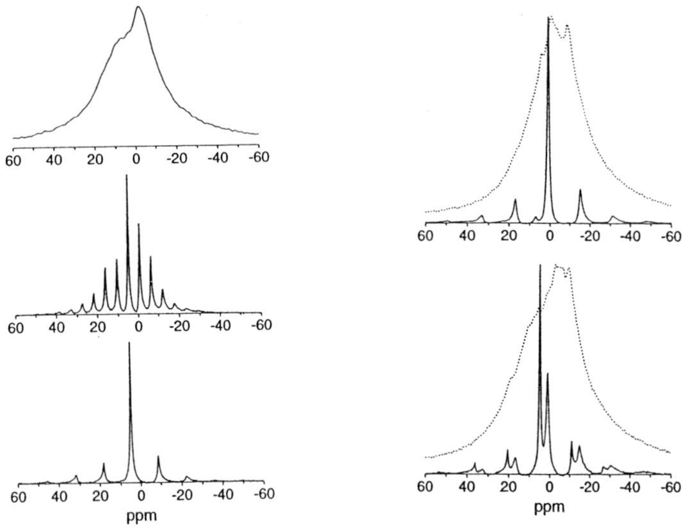

Figure 4.25: The effect of MAS on ^1H NMR spectra of water-saturated Berea sandstone is shown on the left. From top to bottom are spectra of spinning the sample at 0 Hz, 1kHz, and 2.5 kHz, respectively. On the right, it shows the ^1H NMR spectra of Berea sandstone saturated with a light crude oil (top), and a mixture of crude and water (bottom). The rotor spinning speed was 2 kHz in both cases. The dotted line shows the spectra in stationary state, and the solid line shows the spectra at MAS condition. (From de Swiet et al. [71].)

the main Larmor frequency by an exact multiple of the rotor frequency. They will also be stationary in the rotating frame indistinguishable from those with the main Larmor frequency and be collected in the FID. Thus, they will also show up in the Fourier transform of the NMR signal. As the rotor spinning frequency increases, the spinning sidebands move out from the main peak and eventually disappear.

A more interesting aspect of the MAS technique is its application to the wettability issue, as demonstrated by the work of de Swiet et al. [71] shown in the right of Figure 4.25. The top of the figure shows the NMR spectrum of a Berea sandstone, saturated with a light crude of 12 centipoise viscosity, both in stationary state (dotted line) and at a spinning speed of 2 kHz (solid line). The spectrum of the oil shows a large peak in the aliphatic region

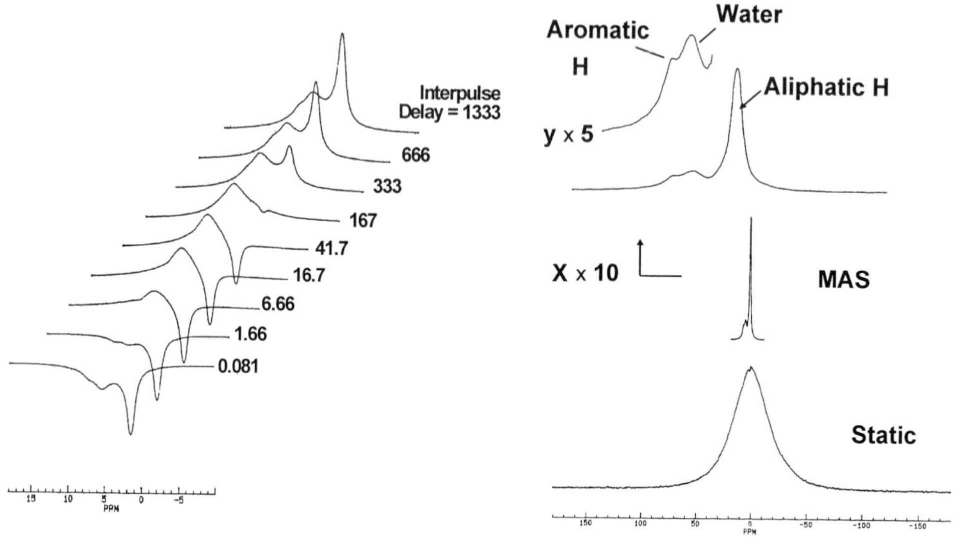

Figure 4.26: The ^{1}H NMR spectra of a T_1 inversion recovery sequence at different τ (ms) in the MAS experiment for a diatomite sample. As shown at the right of the figure, from bottom to top, the ^{1}H NMR spectra of the sample when it is stationary, when it is under the MAS condition, and the enlarged plots (x×10 and then y×5) of the observed peak in the MAS experiment. The major peak is from the aliphatic H, and to the left of the major peak are the water and aromatic H at much smaller amplitudes. (From Wilson and LaTorraca [225].)

and a small peak in the aromatic region (i.e., the tiny bump at the left tail of the central peak), as expected for a light crude. This information is obviously unavailable in the broad peak exhibited by the stationary sample. The bottom of the figure shows the spectrum of a Berea sandstone saturated with a mixture of crude and water. The spectrum exhibits an aliphatic peak and a much stronger water peak which obscures the very small aromatic peak near its base. Assuming the ^{1}H density is the same in oil and water, de Swiet et al. calculated the oil:water ratio within the sample to be $0.95 : 1.0 \pm 0.1$. Considering the sample was prepared by immersion in a mixture of oil:water ratio of 3:1, they concluded the sample was preferentially water wet.

In fact, the wettability issue can be more quantitatively studied by measuring the T_1 relaxation of the sample under the spinning condition. Wilson

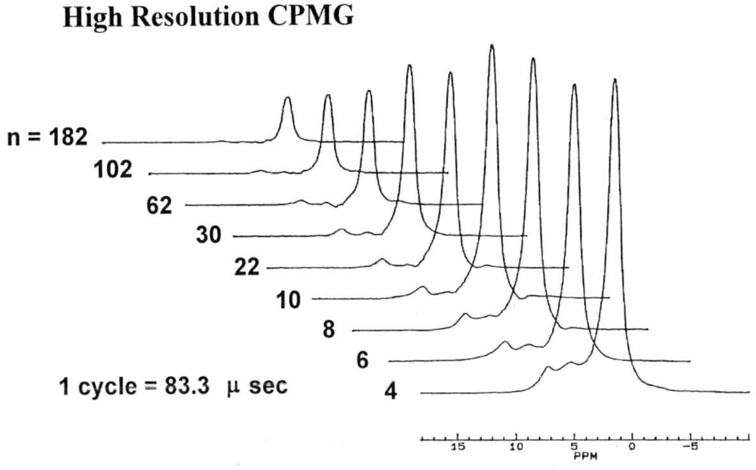

Figure 4.27: The ^{1}H NMR spectra of a T_2 CPMG sequence at selected n-th spin echo in the MAS experiment for a diatomite sample. The major peak is from the aliphatic H, and the minor peaks of water and aromatic H are clearly visible at the left shoulder of the major peak. (From Wilson and LaTorraca [225].)

and LaTorraca [225] show measurements of T_1 (Figure 4.26) and T_2 (Figure 4.27) for diatomite samples under the MAS condition. The high resolution T_1 method is the common inversion recovery sequence with a Fourier transform to yield the spectra. The CPMG sequence was obtained with π pulses synchronized with MAS spinning, i.e., the echo spacing being the same as the spinning cycle, with a Fourier transform to yield the spectrum at selected n-th spin echoes. As shown in Figure 4.27, the major peak is from the aliphatic H, and to the left of the major peak are the water and aromatic H at much smaller amplitudes. The amplitude of each peak can be easily tracked, thus its relaxation behavior can be analyzed independently.

MAS NMR has been shown so far to be a promising technique for core analysis. The mechanical effect of MAS was qualitatively investigated by Xiao et al. [228]. The centrifugal forces may overcome capillary pressure and cause pore fluid to slosh around creating undesirable artefacts for the experiments. Further work is needed so that proper sample preparation procedures can be developed to minimize such an effect.

4.5 NMR core imaging

NMR imaging is mainly used for core imaging in a similar way as the X-ray CAT (Computer-Aided Tomography) scans are used in core studies. The purpose is to understand the fluid distributions within a core before, during, and after a core-flood experiment. The X-ray CAT scan image is based on the contrast in X-ray absorption in different parts of the core. The images depend largely on the high X-ray absorption of the minerals in the solid matrix, not that of the pore fluids. In contrast, all the signal in NMR images comes from the hydrogen nuclei in water or oil in the pore space. The matrix does not contribute to the signal at all. The two techniques are thus complementary.

Unlike the petrophysical applications of other NMR properties, which have a close relationship to rock properties such as irreducible water saturation, pore size distribution, and permeability, NMR core imaging is at present mostly limited to studies of fluid distributions within the pore space. In the future, NMR core imaging may very well develop into a tool for wettability studies and other applications. Because of its importance in core study, as well as in biology and medicine, we shall briefly outline its principles here. For a detailed account, the readers should refer to Callaghan [49].

NMR imaging can perhaps be best understood by first considering an example of one-dimensional imaging: In a T_2 measurement, where the applied field H_0 lies along z and the Larmor angular frequency is $\omega_0 = \gamma H_0$, the magnetization vector is first tipped from the z axis into the xy plane. We then turn on a constant field gradient G_z in the z direction (i.e., the gradient is in the same direction as the spin polarizing field itself) which is so large that it overwhelms the small field inhomogeneities of the instrument and forces different spins to precess differently according to their z coordinate: Each spin precesses at some frequency ω, which differs from ω_0 by

$$\Delta\omega \equiv \omega - \omega_0 = \gamma G_z z. \tag{4.48}$$

Note that this offset can be either positive or negative, since each of the contributing factors G_z, z can be either positive or negative (the gyromagnetic ratio γ for proton is, of course, positive). If the constant field gradient G_z is applied for a time t, those spins will acquire an extra phase $\gamma G_z z t$. Thus, the free induction decay (FID) signal obtained from the sample, while the gradient G_z is on, is given by

$$S(t) = \int \rho(z) \exp(i\gamma G_z z t) dz. \tag{4.49}$$

The inverse Fourier transform of $S(t)$ is then proportional to the one-dimensional proton density profile $\rho(z)$ along the z axis:

$$\rho(z) \sim \int S(t) \exp(-i\gamma G_z z t) dt. \tag{4.50}$$

Extending this observation to three dimensions, when the pulse field gradient $\mathbf{G}$ is a three dimensional vector which is turned on for a time t and can point in *any direction*, the spins at $\mathbf{r}$ now precess at an excess frequency of

$$\Delta\omega = \omega - \omega_0 = \gamma \mathbf{G} \cdot \mathbf{r}. \tag{4.51}$$

Thus, the signal from all $\rho(\mathbf{r})$ is given by

$$S(t) = \int\int\int \rho(\mathbf{r}) \exp(i\gamma \mathbf{G} \cdot \mathbf{r} t) d\mathbf{r}. \tag{4.52}$$

From Eq. (4.52), we can define a "reciprocal space vector" $\mathbf{k}$ in three dimensions as

$$\mathbf{k} = \gamma \mathbf{G} t. \tag{4.53}$$

This allows us to introduce the conjugate pair of functions, $S(\mathbf{k})$ and $\rho(\mathbf{r})$, where $S(\mathbf{k})$ is the spatial Fourier transform of the local proton density in the system $\rho(\mathbf{r})$. These two quantities are related by means of Fourier transformations:

$$S(\mathbf{k}) = \int\int\int \rho(\mathbf{r}) e^{i\mathbf{k}\cdot\mathbf{r}} d\mathbf{r}, \tag{4.54}$$

$$\rho(\mathbf{r}) = \int\int\int S(\mathbf{k}) e^{-i\mathbf{k}\cdot\mathbf{r}} d\mathbf{k}. \tag{4.55}$$

This concept of "$\mathbf{k}$" space was first suggested by Mansfield [150, 151, 152, 153]. With this definition, $\mathbf{k}$ can be made to span the entire three-dimensional space by varying $\mathbf{G}$ and t. Thus, we have constructed a scheme for measuring the "signal" $S(\mathbf{k})$ as a function of $\mathbf{k}$. Using the inverse Fourier transformation of Eq. (4.55), $S(\mathbf{k})$ can be processed to produce the NMR image of $\rho(\mathbf{r})$.

In the above discussion of Eq. (4.52), we neglected the T_2 spin relaxation. When this is not justified, special pulsing schemes should be employed in order to obtain a faithful NMR image.

To illustrate how this is typically done in practice, we consider the example outlined in Figure 4.28: First, a 90° rf pulse, with a sinc function ($\sin x/x$) envelope and a mean carrier frequency ω_0, generates a square shaped slice in the frequency domain *while a large field gradient G_z is applied* along z, in the same direction as the mean spin polarizing field $H_0\mathbf{e}_z$. This procedure selects a thin slab or slice of the sample, parallel to the xy plane. The spins in this slice have a narrow range of Larmor precession frequencies centered around ω_0 and are tilted by the rf pulse into the xy plane. The thickness of this slice is determined by the length of the sinc function and the magnitude of G_z. Only the spins in this thin slice are in resonance with the rf pulse and are, therefore, tipped to the xy plane. Some phase spreading of these spins will occur as a result of the large gradient G_z. The field gradient is then turned off, after which a 180° spin inverting rf pulse is applied. Then the *same field gradient G_z* is turned on again, but for a different length of time, in order to refocus the phases of the precessing spins in the selected slice in an optimal fashion. The purpose of the 180° rf pulse and the subsequent pulse field gradient G_z is to make all the precessing spins in the selected slice have nearly the same phase and frequency, in preparation for the subsequent phase and frequency encoding.

At this juncture, a pulse *field gradient G_y* is turned on, along y, for a time t_y. This causes each spin to precess at a rate that depends on its y coordinate. Thus, it acquires a phase which depends on that coordinate. This corresponds to shifting the origin in $\mathbf{k}$ space by an amount $k_y = \gamma G_y t_y$ along y. This is commonly referred to as "phase encoding".

After that, the so-called "*read*" gradient G_x is turned on, along x, and the FID signal is collected. The gradient G_x now causes the spins with the same y coordinate, which used to have the same rate of precession, to be split into groups with different Larmor frequencies, varying in linear fashion with x. This part of the procedure is called "frequency encoding".

While this gradient G_x is left on, the FID signal is collected from the entire system, as a function of the time t_x elapsed since that gradient was turned on. The signal comes from *only those spins* in the original xy slice that were tipped to the xy plane in the first place, all of which have the *same value of z*! Each digitized signal of the FID corresponds to a particular value of t_x, and thus to a particular value of $k_x = \gamma G_x t_x$. Therefore the signal, recorded as function of t_x, represents a trace along the k_x direction for all those k_x values and with a fixed value of k_y. The digitized FID is therefore a sequence of values of $S(k_x, k_y, z)$, with k_x ranging from 0 to $k_{x,max}$ and at fixed values of k_y and z. By repeating this measurement, using different values of G_y and a fixed time interval t_y, we can collect $S(k_x, k_y, z)$ for *all values of* k_x and k_y at *one fixed value of* z. A 2D NMR image can then be obtained by using the 2D version of the inverse Fourier transformation of Eq. (4.55). In order to obtain the full three-dimensional image, the mean carrier frequency of the initial 90° soft (sinc function) rf pulse is varied to select different xy slices in the sample.

Numerous other field gradient and rf pulsing schemes or sequences can be used for obtaining NMR images of $\rho(\mathbf{r})$. The basic idea is always to find an efficient way of measuring $S(k_x, k_y, z)$ at k_x and k_y values which span that entire 2D subspace of $\mathbf{k}$ space and at a fixed value of z. Figure 4.29 shows low resolution NMR images of the fluids in thin slices through rock samples.

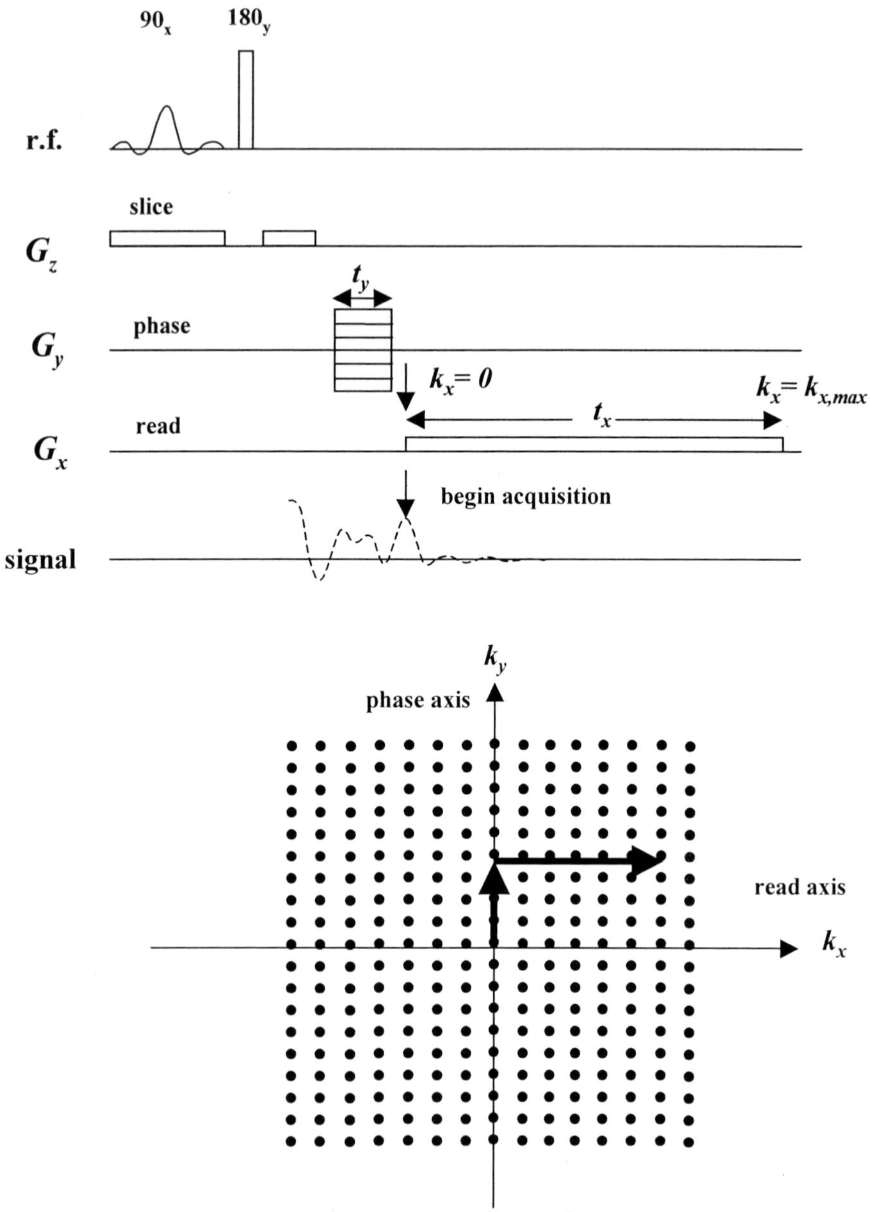

Figure 4.28: The pulsing sequence of getting a 2D NMR image. The bottom of the figure illustrates the path of scanning the **k** vector in **k** space. (From Callaghan [49]. Redrawn from Figure 3.14 of Ref. [49] with author's permission.)

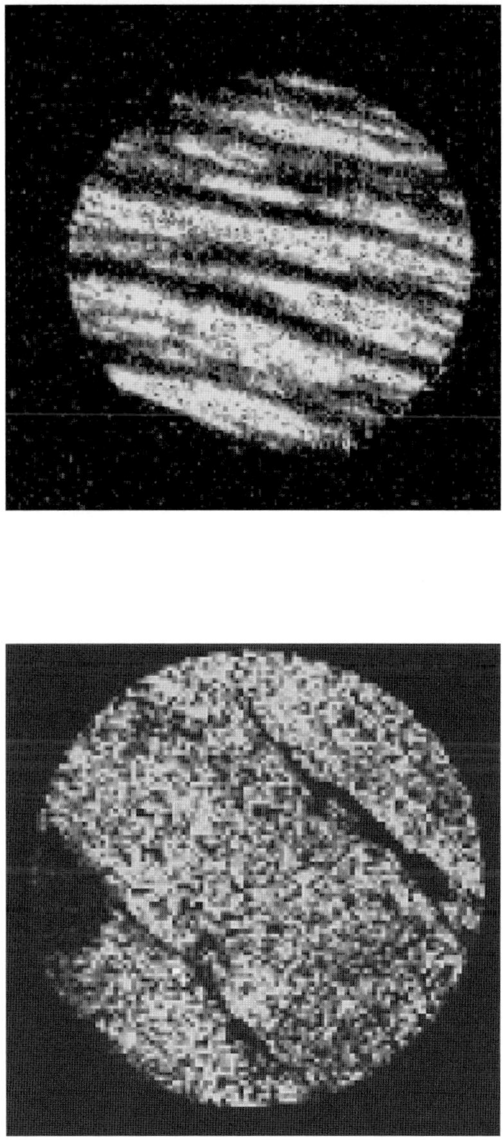

Figure 4.29: NMR images of thin slices through rock samples. The light color represents higher porosity.

Chapter 5

NMR logging applications

There are numerous applications of NMR logging in formation evaluation for the estimation of hydrocarbon reserves and their producibility. As the logging tools improve their capability, more and more new applications of NMR logs are being discovered. The most frequently used applications now are: (1) porosity estimates, (2) irreducible water saturation determination, (3) permeability prediction, (4) residual oil estimation, (5) hydrocarbon typing, and (6) oil viscosity estimation. At the end of this section, we shall also discuss the logging guidelines, namely, some of the issues one should be aware of in the planning and preparation of an NMR logging run.

5.1 Porosity estimation

An NMR logging tool measures the total population of hydrogen nuclei in the formation, including the hydrogen atoms of the water and oil in the pore space, and those in the clay. Hydrogen atoms in the solid matrix, if any, are generally not sensed by the logging tools, due to the extremely long T_1 and short T_2. The current NMR logging tools typically measure hydrogen atoms with relaxation times down to 1 ms, or in some cases, can be as low as 0.3 ms if the early signals are stacked many times. In general, the measured NMR signal only reflects the hydrogen nuclei in the pore space.

Because the measurements are sensitive to proton (i.e., hydrogen nucleus) spins in fluids only[1], the estimated porosity value should be mineralogy in-

[1] There are exceptions. Sometimes, the logging conditions are such that the ^{23}Na signal is also measured [154] inadvertently.

dependent. As the crude oils have a hydrogen index very similar to that of water, porosity can be reliably measured in both water and oil productive zones. This measurement is considered *mineralogy independent* in reference to other porosity logging tools which have a significant response to the mineralogy as well as to pore fluids. The sensitivity of these other logging tools to mineralogy means that additional information or assumptions must be made to obtain porosity. However, there are several factors which could affect the NMR porosity estimates as well. These are discussed in the following.

Ferro- and paramagnetic materials

The presence of some iron bearing minerals or high magnetic susceptibility pore-lining materials can greatly reduce the T_2 relaxation times. As a result, some NMR signals with very short T_2 relaxation times are lost due to finite dead time of the instruments. This will lead to reduced porosity values. This situation is often accompanied by a shift of the T_2 peaks toward shorter relaxation times due to the large internal magnetic field gradient induced by the presence of these minerals. Thus, the NMR porosity determination is mineralogy independent, except when highly paramagnetic minerals are present.

Tool calibration

Improper tool calibration can lead to erroneous porosity readings. The measured NMR signal is calibrated against a known standard (typically a water sample) to give values in porosity units. The initial signal amplitude of a T_2 CPMG measurement (i.e., extrapolating the echo train back to zero time) is proportional to the density of hydrogen nuclei in the probed zone. Since this quantity depends on a number of factors, such as the rate of signal decay, the noise level of the measurement, the loading on the rf coil, etc., care should be taken to ensure it is properly done. Errors arising from tool calibration mistakes can often be easily identified because they affect the porosity value of the whole log.

Environmental corrections

There are a number of environmental corrections that need to be made before the measured NMR signal can yield the correct porosity values. First of all, the NMR amplitudes should be corrected for the temperature and field strength to reflect their proportionality to the hydrogen spin density. From Curie's law:

$$M = N\gamma^2\hbar^2 I(I+1)H_0/3kT \tag{5.1}$$

where

$$\begin{aligned} M &= \text{magnetization} \\ N &= \text{spin density} \\ \gamma &= \text{gyromagnetic ratio} \\ \hbar &= \text{Planck constant}/2\pi \\ I &= \text{spin quantum number} \\ H_0 &= \text{magnetic field} \\ k &= \text{Boltzmann's constant} \\ T &= \text{absolute temperature} \end{aligned}$$

Thus, the magnetization, M, which is proportional to the NMR amplitude, increases with field strength, and decreases with increasing temperature, for the same amount of water volume. Hence, it must be corrected by a factor T/H_0, i.e., the formation temperature and the logging frequency. The temperature and salinity also affect the hydrogen density of water. In addition to this, the salinity, thus the conductivity of the water, can affect the Q of the rf coil and reduce the signal strength. All of these are considered in the logging tool design and the signal value conversion, and are normally properly corrected by the logging companies. If some of these environmental corrections are inadvertently left out, or if special circumstances lead to improper corrections, the measured porosity values will be affected.

Insufficient polarization time

Under certain logging conditions (e.g., high temperature/long T_1 of pore fluids, vuggy carbonates, low surface relaxivities, ...), repeat delay times greater than 10 seconds are often required. Insufficient wait time between successive excitation pulses can cause incomplete recovery of the spins to their equilibrium state. This will lead to a reduction of the measured signal, and thus of the porosity, by the following factor

$$a = 1 - e^{-RD/T_1} \tag{5.2}$$

where RD stands for the Repeat Delay time (or wait time), i.e., the time between the end of a CPMG echo train and the start of the next CPMG excitation, and T_1 is the spin-lattice relaxation time for the fluid. This factor, a, is commonly known as the polarization factor. Typically, the wait time is chosen to be three times T_1. This ensures that 95% of the spins have been polarized along the applied magnetic field direction.

The loss of porosity due to incomplete recovery can often be corrected by assuming a T_1/T_2 ratio and including the polarization factor in the data inversion process. Since, the T_1/T_2 ratio is not always a constant for the formation fluids for the whole logged depth, this exercise is sometimes not successful. This is especially true when there are multiple formation fluids such as water and oil, which can have very different T_1/T_2 ratios.

Presence of gas

The presence of free gas or gas dissolved in oil can also lead to errors in porosity. The hydrogen index (HI = the ratio of the signal from a volume of gas or oil to the signal from an equal volume of water) is significantly less than 1 even for very high pressure gas. Unless accounted for, this will cause the porosity to be underestimated.

Gas dissolved in oil reduces the viscosity of the oil without significantly increasing the oil volume. The viscosity reduction increases both T_2 and T_1. These increases mean that longer wait times are needed with the potential for error during tool setup if the wait time is determined in a wet zone. The presence of gas in the oil may also decrease the HI of the oil which would lead to underestimation of porosity (see discussion in Chapter 3).

Presence of clay, heavy oils, or bitumen

When considering the contributions of clay to porosity, we touch upon an area of controversy. Some argue that the water of hydration in clays should be included as a part of the porosity, while others argue against it. Without taking sides in this controversy, we shall simply state that NMR measures the formation volume containing hydrogen nuclei with T_2 relaxation times greater than 1~3 ms, depending on the noise level. If a higher precision of measurement is achieved, the lower limit on T_2 relaxation time can be extended to 0.3 ms or less.

As discussed previously [177], current literature suggests that the T_2 of clays covers a wide range of relaxation times from less than 0.1 to over 16 ms. Thus, in high precision measurements, most if not all of the porosity associated with clay will be included.

The T_2 relaxation times for heavy oils vary over a wide range depending on their viscosity. For viscous high density oils, the NMR log cannot capture all the signal and may suffer an apparent loss of porosity.

The T_2 relaxation time for solid bitumen is very short, often around 0.1 ms or less. Thus, solid bitumen cannot be measured with the current NMR logging tools. Because the neutron log measures bitumen and brine-saturated pore space, the combined information of neutron and NMR porosities may allow us to estimate the bitumen content of the formation.

Other factors

Additional sources of error in porosity are tool related. High well bore temperatures can cause tool malfunctioning. Cross talk between other wireline tools mounted on the same string can cause an erratic signal or an unacceptable noise level.

5.2 Irreducible water saturation

One of the important applications of NMR logging is estimating the irreducible water saturation. This is the portion of the capillary-bound water plus the clay-bound water in the pore space. Under a certain capillary pressure, this part of the water in the formation is not producible, hence the term, irreducible water. The knowledge of this information has at least two impacts: (1) determining the water cut, i.e., the amount of water to be produced from the reservoir, and (2) estimating the permeability by using the Coates equation as discussed in Chapter 4.

The irreducible water volume can be determined from NMR measurements/logs as discussed in Chapter 4. For using a T_2 cutoff to determine the irreducible water saturation, a T_2 relaxation time of 33 ms is typically chosen for sandstones, and 100 ms is used for carbonates. As discussed previously, these are values obtained based on a statistical average from experimental measurements. Different reservoirs may have very different rock fabrics with different capillary pressure behaviors. It is recommended that NMR core measurements be performed for the reservoir so that a more specific T_2 cutoff, or a spectral method [64] can be applied.

To determine the water cut, or the producible water, one can subtract the irreducible water volume from the water volume estimated from the resistivity log. If the water saturation determined from the resistivity log, S_w, is significantly greater than the irreducible water saturation estimated from the NMR log, S_{wirr}, the reservoir will produce water early in its productive life. If $S_w \cong S_{wirr}$, the reservoir is at irreducible water saturation and no water will be produced in its productive life. In principle, the formation cannot have a water volume which is less than the irreducible water measured by NMR. However, such a situation is possible if the capillary pressure of the reservoir is higher than the air/water capillary pressure used in the laboratory for determining the NMR T_2 cutoff for irreducible water saturation.

The T_2 cutoff for sandstones appears to work best in well consolidated formations with permeabilities in the 0.3 md - 3,000 md range where the wells are drilled with water-based mud.

For poorly consolidated formations, reservoirs with long oil columns, formations with significant amounts of ferro- and paramagnetic compounds (e.g., iron, manganese), and reservoirs with heavy oil or bitumen, specific T_2 cutoffs may have to be established by core measurements.

Poorly consolidated clastics often have very low irreducible brine saturations in the range of 5-20%. For these rocks, the amount of T_2 signal below 33 ms is small and can be hard to resolve from the noise. Reservoirs with long oil columns may require cutoffs lower than 33 ms. Significant amounts of paramagnetic compounds on the grain surfaces can lead to large induced magnetic field gradients which shift the T_2 distributions to short times. Reservoirs with heavy oil can have T_2 distributions with overlapping oil and brine signals such that the irreducible water saturation, S_{wirr}, can not be readily resolved. Partially oil-wet reservoirs can also have overlapping T_2 distributions for the oil and brine.

The presence of heavy oil or bitumen, which has very short T_2 relaxation times, further complicates the issue. When one interprets the NMR T_2 distributions, one has to keep all these in mind.

Sources of T_2 values shorter than 3 ms include: water in extremely fine grain silts or clays, very heavy oils or bitumen, paramagnetic compounds in solution or on the grain surfaces, and hydrated water in gypsum or clay minerals.

5.3 Permeability prediction

As discussed in Chapter 4, there are two commonly used permeability estimators, i.e., the Coates/Timur equation[63, 214] and the $T_{2,LM}$ equation [121, 11]. Sezginer et al. [199] recently suggested a method that uses the summation of echoes to predict the permeability with some success.

Coates/Timur equation The Coates equation uses porosity, BVI, and FFI to predict permeability. Therefore, whenever the BVI and porosity can be accurately determined, this is the preferred method. However, the predicted value is quite sensitive to the value of BVI, and it generally tends to overestimate permeability.

$\mathrm{T}_{2,\mathrm{LM}}$ **equation** The $T_{2,LM}$ equation uses the logarithmic mean of T_2 distribution to predict permeability. The estimator is derived based on brine-saturated sedimentary rock samples. It generally gives good prediction for the water zone. However, in the oil zone, since the T_2 relaxation times for oils are in general different from those for water, the $T_{2,LM}$ equation is not reliable.

Summation of echoes The summation of echoes is related to the mean relaxation time, $\langle T_2 \rangle$, hence the surface to volume ratio of the large pores of the rock formation, and hence the permeability. Although laboratory experiments were carried out on a few quarried sandstone samples to show such a correlation [199], not enough laboratory measurements were performed to establish a more general and robust formula. It can be used as a qualitative indicator for permeability. For a more quantitative estimation, it is recommended that field specific formulas be established through NMR core measurements. One important advantage of this method is that it eliminates the noise by summing echoes in the time domain, thus retaining the high resolution of NMR information from pad-type tools such as Schlumberger's CMR without the need to stack the signals depth-wise.

With these three types of permeability estimators, we should be selective in properly applying them. As different reservoir conditions dictate different considerations in the application of these permeability estimators, we shall discuss a few as follows:

Water based mud, highly invaded reservoir

When water-based drilling muds are used and the formation is extensively invaded by mud filtrate, the $T_{2,LM}$ based method may be preferred because the $T_{2,LM}$ estimate is less sensitive to noise than is the BVI determination. This assumes that the reservoir is water-wet, and that all of the hydrocarbon has been flushed from near the well bore.

Water based mud, significant remaining low viscosity oil

When significant amounts of oil remain in the formation, the T_2 distribution may look bimodal, with the long time associated with oil where the oil has low viscosity and the reservoir is water-wet. Here, the Coates equation, or the $T_{2,LM}$ equation modified to account for the presence of oil, may be preferred.

Water based mud, significant remaining high viscosity oil

Neither equation will work as written when the oil and brine T_2 distributions overlap to the extent that they cannot be distinguished. Here, the oil and brine signals overlap, and the oil, though non-wetting, has a broad distribution of T_2 values. This is a difficult situation. No straightforward satisfactory solution has been found. One of the possible solutions is to combine diffusion measurements (discussed below), resistivity log and laboratory measurements of oil to determine the irreducible water saturation.

To illustrate the overlap between the oil and brine signals, the T_2 distributions for a reservoir rock sample are shown in Figure 5.1 before and after soaking in D_2O. The first measurement has both signals from the oil and brine whereas the second measurement has only the oil signal.

Oil based mud, low viscosity invading filtrate

When a well is drilled with oil based mud containing low viscosity oil filtrate, the Coates equation is preferred especially where the invading filtrate displaces the reservoir hydrocarbon. An example of this case is when the light oil filtrate displaces gas and results in a bimodal T_2 distribution where the brine and oil peaks are clearly separated. Irreducible brine saturation is determined directly from the area under the short time peak, and accurate estimates of permeability are obtained. Stonard et al. [206] used this approach to obtain the transform shown in Figure 5.2. Note that the transform

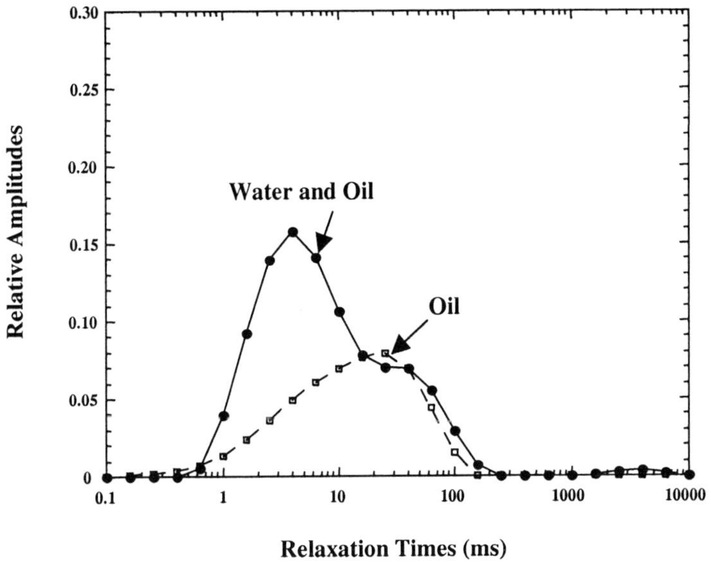

Figure 5.1: The T_2 distribution of a rock sample before and after soaking in D_2O to remove the water signal.

compares favorably with the core permeabilities through almost all of the log interval.

Field specific permeability algorithms

Even when the correct algorithm is chosen for a given field, the estimated permeabilities may be way off because of local conditions. The permeability algorithms supplied by the logging companies were based on laboratory measurements on a finite range of clastic reservoirs where, for example, the permeabilities were typically greater than 20 md. The laboratory based transform for the permeabilities of core plugs from various reservoirs can be very different from the commonly used transform $k = \phi^4 T_{2,LM}^2$.

Carbonate reservoirs

For carbonate reservoirs, it has long been known that NMR log derived permeability estimates are not as satisfactory and consistent as those for clastic sediments. One of the reasons for such difficulty is that there are intergranular and vuggy porosities. Chang et al. [56] found that if the porosity above

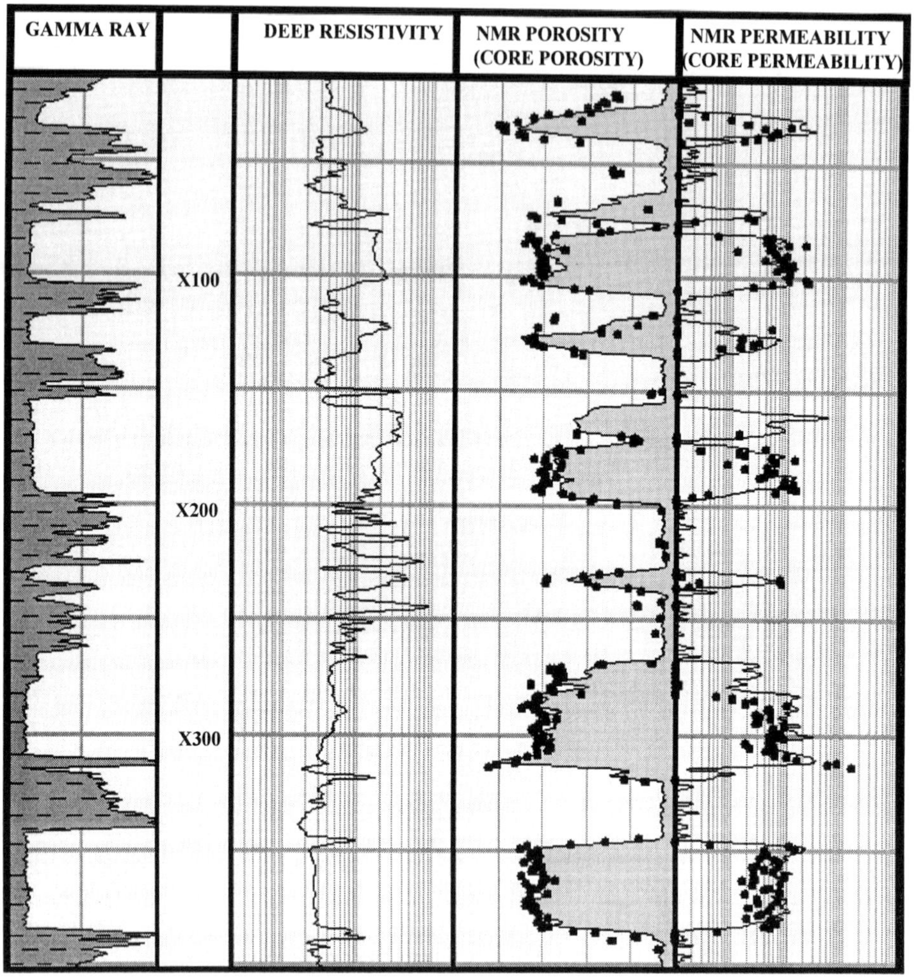

Figure 5.2: The log data showing NMR predicted permeability (curves) compares favorably with core permeabilities (solid data points). The comparison of NMR porosity estimates (curves) with core porosities (solid data points) is also shown. (From Stonard et al. [206].)

T_2 relaxation time of 750 ms, presumably the vuggy pores, is excluded, the $T_{2,LM}$ equation for permeability prediction works quite well. Chang et al. came up with a permeability formula based on laboratory NMR core measurements, i.e.,

$$k = 4.75(\phi_{NMR,<750ms})^4(T_{2,<750ms})^2. \tag{5.3}$$

However, one should be aware that laboratory determined vuggy porosity T_2 cutoff may need to be modified when applied to NMR log analysis. The logging tools all have significant field gradients which will yield an apparent T_2 which in general, because of the diffusion effect, is shorter than the T_2 determined in the laboratory where the field is normally uniform.

5.4 Residual oil determination

Horkowitz et al. [103] provided a good account on using NMR logging for residual oil determination. Prior to the introduction of the current pulsed NMR logging technology, Robinson et al. [188, 189] and Setser and Williams [198] already applied the NMR inject-log technique for the residual oil saturation determination in open holes. They used dopants such as Mn-EDTA (Ethylene-Diamine-Tetra-Acetic acid) in the injected brine to kill the water signal. The presence of Mn^{++} ions shortens the relaxation time of water to below the dead time of the earth's field tool (about 30 ms[2]) so that only the oil signal is detected.

The advent of the new generation of pulsed NMR logging tools has greatly simplified this method. First of all, there is no need to dope the drilling mud with magnetite to kill the borehole fluid signal. Second, the reduction of the tool dead time from 30 ms to about 1 ms allows the Mn-EDTA doping procedure to measure the total porosity as well as the oil saturation. The Mn-EDTA doping provides the Mn^{++} concentration in the pore fluid needed to shorten the relaxation time of the water, to such an extent that the water signal is completely separated from the oil signal, but not too short, so that all the water signal is above the dead time and is still measurable by the tool. Figure 5.3 shows schematically such a procedure.

Paramagnetic ions dissolved in water, such as Mn^{++} and Gd^{++}, shorten the relaxation time of the hydrogen nuclei of water by the translational

[2] The dead time for the experimental NML was about 30 ms. The final Schlumberger tools had an effective dead time of 21 ms.

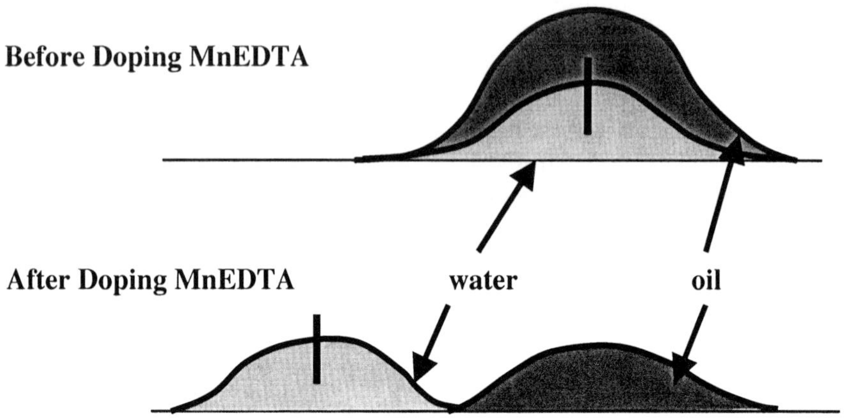

Figure 5.3: The T_2 distributions before and after Mn-EDTA doping for residual oil determination.

dipole-dipole interaction between the nuclear magnetic moment of the proton and the electronic magnetic moment of the paramagnetic ions. Bloembergen et al. [24] showed that the proton spin relaxation rate of such a solution of paramagnetic ions is given by

$$\frac{1}{T_{1,2}} = \frac{16}{15}\frac{\pi^2\gamma^2 N_{ion}\mu^2\eta}{kT}, \tag{5.4}$$

where γ is the proton gyromagnetic ratio, N_{ion} the ion concentration, μ the effective magnetic moment of the ion, η the viscosity, k the Boltzmann constant, and T the absolute temperature.

In addition to the dipole-dipole relaxation, the hydrogen nuclei of the water molecules also relax by spin exchange interaction with the Mn^{++} and Gd^{++} ions (Bloembergen [25]). For Mn^{++}, the spin exchange interaction is about seven times stronger than the dipole-dipole interaction.

The manganous ion is often selected for doping the drilling mud because of its high magnetic moment, good solubility, and availability. A chelating agent (such as EDTA) is frequently used to prevent Mn^{++} from forming $Mn(OH)_2$ precipitates. Thus chelation suppresses the spin exchange interaction between Mn^{++} and water molecules, and reduces the effect of Mn^{++} in shortening the relaxation time. Chelation also prevents ion exchange on charged surfaces such as clays in drilling mud or reservoir. As a general rule, in shaly sands or bentonite drilling muds, the Mn^{++} dopant should be chelated with EDTA to prevent flocculating the clays.

As described by Horkowitz et al. [103], the paramagnetic solution of Mn-EDTA can be prepared by mixing a 55-gal drum of Versene 100 liquid (38% Na_4EDTA) with 100 lbs of anhydrous manganous sulfate per 50 barrels of water. This results in a solution of 0.03M in manganese containing 15% excess EDTA. An excess of EDTA compared to Mn^{++}, typically 10% to 30%, is required to prevent the precipitation of 2Mn-EDTA.

Using Mn^{++} dopant in NMR logging to determine residual oil saturation apparently has been successfully applied in many different parts of the world (Horkowitz et al. [103], Crowe et al. [70], Hassoun et al. [98]). Horkowitz et al. [103] suggested using $MnCl_2$ instead of Mn-EDTA, with many added benefits besides reducing costs, in carbonate reservoirs where clay flocculation is not a concern.

Estimating residual oil saturation in wells drilled with oil based mud has been considered difficult or impossible. This is particularly true when using T_2 measurements to distinguish mud filtrate and low viscosity native formation oil whose T_2 contrast is small. Flaum et al. [83] suggested a fast logging procedure to provide a signal contrast when the T_1 contrast between mud filtrate and formation oil is large. Using this method, they were able to estimate residual oil saturation in the flushed zone successfully.

5.5 Hydrocarbon typing

Using NMR log to perform hydrocarbon typing was first suggested by Akkurt et al. [3], primarily for the NMR logging tools which can provide a constant field gradient, such as Halliburton/Numar's MRIL. Basically, they are using two logging parameters to obtain a contrast in the NMR signals for water, oil, and gas, i.e., the repeat delay time (variable RD) and the echo spacing (variable TE). From the analyses of the NMR data, at least qualitative hydrocarbon typing, and sometimes quantitative hydrocarbon estimates, are possible.

The method of variable TE requires a simple, well-defined field gradient. Thus, it is often limited to Halliburton/Numar's MRIL. The method of variable RD, on the other hand, can be applied to both Halliburton/Numar's MRIL as well as to Schlumberger's CMR. In fact, elaborate schemes have been proposed to use the method of variable TE for hydrocarbon typing even for a complex field gradient such as that of Schlumberger's CMR, as we shall discuss later in this chapter.

The combination of variable TE and variable RD can be used not only for hydrocarbon typing but also for oil viscosity estimation [58, 59]. Recent work by Freedman et al. [90] has shown that with variable TE, variable RD, and variable field gradient, one can identify the water, oil, and gas phases, both qualitatively and quantitatively, and even provide an estimate for the oil viscosity. This is discussed in Section 5.6.

5.5.1 Multiple repeat delay times

In a standard CPMG run, a quantity called repeat delay time (or wait time, time to repeat, or RD, WT, TR for short) has to be selected. This is the length of time between any two successive CPMG echo trains. This time should be chosen to be long enough so that all spins that are tipped from the z axis into the xy plane have sufficient time to return to their original thermodynamic equilibrium state. This time is typically chosen to be 3 to 5 times the maximum T_1 relaxation time. If the RD is not long enough, then some of the tipped spins may not have fully returned to their thermal equilibrium state and a reduction of the total signal will occur, with the reduction factor given by:

$$a = 1 - e^{-RD/T_1} \tag{5.5}$$

When RD is 3 to 5 times of T_1, the factor a, commonly known as *polarization factor*, is very close to one. If RD is smaller than T_1, the factor a can be much less than one.

Note that both gas and oil are typically non-wetting phases. Their T_1's can be several seconds, which are much longer than the T_1 for water in the rock (typically several hundred milliseconds). As discussed in Chapter 3, while the apparent T_2 of the oil is only slightly shifted by the field gradient, the T_2 for gas is significantly shifted to shorter relaxation times in the same gradient field due to its large diffusion coefficient. An example of T_1 and T_2 for reservoir water, oil, and gas in a field gradient of 17 Gauss/cm with a TE $(= 2\tau)$ of 1.2 ms is shown in Table 5.1.

Table 5.1: An example of T_1 and T_2 of reservoir fluids in a field gradient of 17 Gauss/cm for a typical sandstone reservoir, where the TE ($=\tau$) is 1.2 ms.

	T_1 ms	D 10^{-5}cm^2/s	$\frac{1}{3}\gamma^2 G^2 D\tau^2$ s^{-1}	$T_{2,apparent}$ ms
Brine	1-500	7.7	1.9	0.67-200
Oil	1000	2	0.5	670
Gas	4000	100	24.8	40

Thus the T_2 distribution for the reservoir rock containing water, oil, and gas could have a general form as illustrated schematically in Figure 5.4. One can have two logging passes with a RD_S shorter than $T_{1,gas}$ for gas, or $T_{1,oil}$ for oil, but long enough for the longest T_{1w} component of the water in the rock to recover, and another RD_L long enough for capturing all gas and oil signals as follows:

$$3T_{1w,\max} < RD_S < T_{1,gas} \text{ (and/or } T_{1,oil}) < RD_L. \tag{5.6}$$

Then by subtracting the pass with the short wait time, RD_S, from that with the long wait time, RD_L, the portions for the water signal cancel out, leaving only the contributions from the gas and oil, as shown in the following:

$$\phi_{RD_L} = \phi_w + \phi_{gas}(1 - e^{-RD_L/T_{1,gas}}) + \phi_{oil}(1 - e^{-RD_L/T_{1,oil}}) \tag{5.7}$$

$$\phi_{RD_S} = \phi_w + \phi_{gas}(1 - e^{-RD_S/T_{1,gas}}) + \phi_{oil}(1 - e^{-RD_S/T_{1,oil}}) \tag{5.8}$$

$$\phi_{RD_L} - \phi_{RD_S} = \phi_{gas}e^{-RD_S/T_{1,gas}} + \phi_{oil}e^{-RD_S/T_{1,oil}} \tag{5.9}$$

where ϕ_{RD_L} and ϕ_{RD_S} represent the T_2 signals for long and short repeat delay times, respectively, and ϕ_w the water signal, ϕ_{gas} the gas signal, and ϕ_{oil} the oil signal. The factors, $e^{-RD_L/T_{1,gas}}$ and $e^{-RD_L/T_{1,oil}}$, presumably are very small and can be dropped out.

From Eq. (5.9), one should be able to obtain the gas and oil saturations, assuming the hydrogen index for the gas is known. This is the basis for

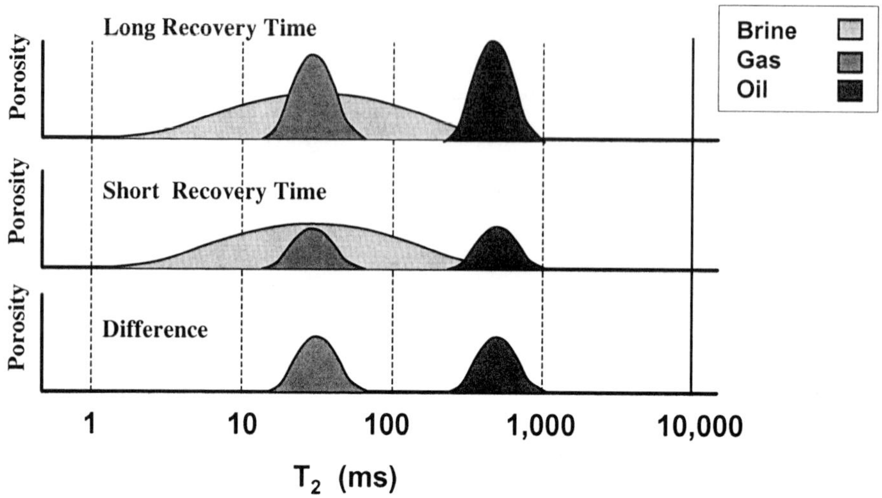

Figure 5.4: Schematic diagram of T_2 distributions for a reservoir rock containing water, oil, and gas in a differential spectrum procedure.

the multiple repeat delay time method, also known as Differential Spectrum Method (DSM).

In practice, subtracting the T_2 distribution of the short repeat delay time from that of the long repeat delay time usually does not work very well. This is because the typical noise level in the NMR logging measurements is quite high, and there are uncertainties of the same order of magnitude as the answer we are seeking in the T_2 distributions.

Prammer et al. [178] extended Akkurt et al.'s method [3] by doing this in the time domain, i.e., subtracting the echo train of the short repeat delay time from that of the long repeat delay time, and inverting the difference of the echo trains to obtain the T_2 distribution. This is called the *time domain analysis* (TDA). This approach is more reliable than the subtraction in the T_2 domain, because it bypasses the issue of uncertainty associated with the inversion. To obtain reliable results, frequently a significant stacking of the data of many depth intervals is required. In order to have a meaningful outcome, a relatively large volume of high pressure gas is required to have a discernible difference between the two logging passes with different RD's.

5.5.2 Variable echo spacing

There is another parameter in the CPMG T_2 measurements which we can manipulate to extract more information, i.e., the echo spacing, or the time between echoes TE. As described in Chapter 2, the T_2 relaxation rate of bulk water in a constant field gradient is modified as follows:

$$\text{Rate} = \frac{1}{T_{2B,w}} + \frac{1}{3}\gamma^2 G^2 D_w \tau^2, \tag{5.10}$$

where $T_{2B,w}$ is the T_2 for bulk water, G the constant field gradient, and D_w the self diffusion coefficient of water. For porous media such as rocks in a constant field gradient, Eq. (5.10) is further modified as:

$$\text{Rate} = \frac{1}{T_{2B,w}} + \frac{1}{T_2} + \frac{1}{3}\gamma^2 G^2 D_w \tau^2, \tag{5.11}$$

where second term is related to the surface to volume ratio:

$$\frac{1}{T_2} = \rho \frac{S}{V}, \tag{5.12}$$

and the first term is negligible compared to the second term. Thus, when we have a fluid-saturated rock with a wide distribution of pore sizes in a constant field gradient which is larger than internal gradients from susceptibility differences, at times short enough that most fluid molecules have not hit pore walls or phase boundaries, we would expect each relaxation rate $1/T_{2j}$ to increase by an amount of

$$\frac{1}{3}\gamma^2 G^2 D_w \tau^2, \tag{5.13}$$

where τ=TE/2. Hence, by increasing TE, we would expect the T_2 amplitude distributions to shift to shorter relaxation times. Since the increase in the relaxation rate is a constant, the shift of T_2 due to this diffusion term is larger at long relaxation times but quite small at short relaxation times. For example, if the diffusion term $\gamma^2 G^2 D_w \tau^2$ is $2s^{-1}$, then the relaxation time at 500 ms would shift to 250 ms, whereas the relaxation time at 5 ms would only shift to 4.95 ms.

Because oil is often the non-wetting phase, its relaxation rate, or $1/T_{2,o}^{\dagger}$ (where $T_{2,o}^{\dagger}$ is the apparent $T_{2,o}$ value in the constant field gradient, and we

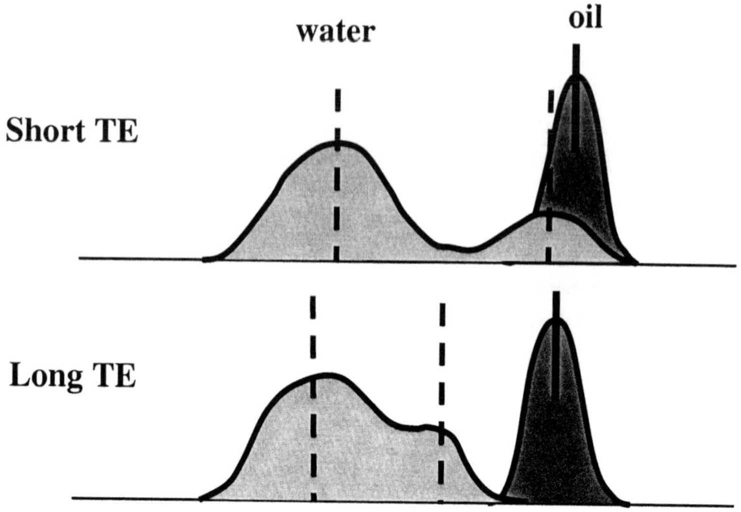

Figure 5.5: In the shift spectrum scheme, the T_2 distribution of a long TE helps resolve water and oil signals that were overlapping in the T_2 distribution of a short TE run.

use the subscript "o" to represent oil), depends on the bulk $T_{2B,o}$, its diffusion coefficients D_o, the gradient strength G, and the half echo spacing τ, i.e.;

$$\frac{1}{T_{2,o}^{\dagger}} = \frac{1}{T_{2B,o}} + \frac{1}{3}\gamma^2 G^2 D_o \tau^2 \tag{5.14}$$

The diffusion coefficient of oil, D_o, is usually an order of magnitude less than that of water. Thus, the diffusion term in the relaxation rate equation (Eqs. (5.11) and (5.14)) for water is much larger than that for oil. As a result, as the TE is increased, the shift of T_2 for water is much more than that for oil. Therefore, if the signals for water and oil overlap initially, a larger TE can often produce enough relative shift in T_2 and hence resolve the water and the oil signal. This is usually quite effective for long T_2 relaxation times (several hundred milliseconds), because the shift there is quite significant. Figure 5.5 shows schematically such a shift in T_2 peaks of water which helps resolve the oil signal.

The diffusion coefficient of gas is about two orders of magnitude larger than that of water. The apparent T_2 for gas for a field gradient of 17

Gauss/cm and a TE of 1.2 ms is already in the neighborhood of 40-60 ms. Further increase in TE would shift the gas signal into the region of irreducible water. Besides, the gas signal is very weak, and it is very hard to detect the effect of varying TE. Hence, the *Shift Spectrum Method* is commonly used for detecting oil, whereas the *Differential Spectrum Method* can be used for both gas and oil.

The variable TE method or the *Shift Spectrum Method* has been used by many to extract oil saturation from NMR logs. Looyestijn [145] used the large contrast of diffusion coefficients between water and oil and NMR logs taken with different echo spacings to solve for the oil saturation with reasonable success. Slijkerman et al. [203] used forward modeling of a multifluid relaxation model to solve for oil and water saturations.

5.5.3 Enhanced diffusion

The method of variable wait time relies on a large T_1 contrast between hydrocarbon and water. When such contrast becomes small, e.g., when the T_1 of the oil is close to that of the water, this method is not effective. If however, the diffusion coefficient of the oil is still reasonably larger than that of the water, then with a large enough TE, enough contrast in the T_2 for oil and water can still be produced. Based on this, Akkurt et al. suggested [4, 5] a single large TE (4.8 or 6 ms) for a dual wait time run to resolve this problem. They called this *Enhanced Diffusion Method* (EDM). Figure 5.6 shows schematically the general concept of the procedure.

Recall that the apparent T_2 for water in a gradient field is given by

$$\begin{aligned} \frac{1}{T_2^\dagger} &= \frac{1}{T_{2B,w}} + \frac{1}{T_{2,surf}} + \frac{1}{3}\gamma^2 G^2 D_w \tau^2 \\ &= \frac{1}{T_{2,int}} + \frac{1}{T_{2D}} \end{aligned} \tag{5.15}$$

where the first term on the right hand side, $1/T_{2B,w}$ (the bulk relaxation rate), can often be neglected; the second term, $1/T_{2,surf}$ (the surface relaxation rate), when combined with the first term, can be designated as the intrinsic relaxation rate, $1/T_{2,int}$; and the last term, $1/T_{2D}$, is defined as

$$T_{2D} = 3(\gamma^2 G^2 D_w \tau^2)^{-1} \tag{5.16}$$

A key point, brought forward by Akkurt et al. [4, 5], is that T_{2D} is the upper bound of all the water signal. This can be easily appreciated by consid-

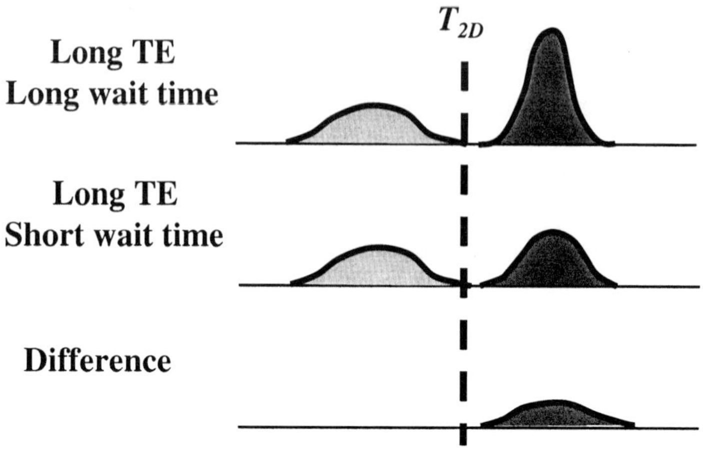

Figure 5.6: The T_2 distribution of long TE for dual wait times can help remove the water signal in an enhanced diffusion run. The T_{2D} is the upper bound for the water signal.

ering Eq. (5.15). For all the possible values that $T_{2,int}$ can have, the apparent $T_2^\dagger$ will always be less than T_{2D}. If $T_{2,int}$ is small, then $T_2^\dagger \approx T_{2,int}$. When $T_{2,int}$ tends to infinity, then $T_2^\dagger \approx T_{2D}$, with T_{2D} being its upper limit. Thus, when a large TE is used, T_{2D} shifts toward shorter relaxation times. Any signal beyond T_{2D} must then come from oil.

This becomes a simple way to judge the nature of the signal beyond T_{2D}, i.e., it must come from oil. However, it does not preclude the possibility that part of the signal below T_{2D} can also come from oil, or that all oil signal happens to be less than T_{2D}. Thus, Enhanced Diffusion Method (EDM) allows you to make a statement about some of the observations, but does not provide you with a complete picture unless you have more information, such as laboratory NMR measurements of the produced oil of the reservoir. Other issues such as the noise level of the signal can also cloud the judgment about T_2 components beyond T_{2D}. The regularization process[3] of noisy data can often cause water signal to spill over the T_{2D} limit.

[3]Regularization is a technique of NMR data inversion which is discussed in Chapter 6.

With the laboratory information of the produced oil, as well as the dual wait time data set, it is possible to perform elaborate time domain analysis to extract a more accurate estimate of the oil saturation.

5.5.4 Complex field gradients

Our discussion so far on the variable RD and TE methods are mostly used for NMR tools which have a simple, single-valued, and well-defined constant field gradient, such as Halliburton/Numar's MRIL. For Schlumberger's CMR, which has a complicated field gradient distribution, implementation of these methods is not as straightforward. Generally speaking, the variable RD method requires a good T_1 contrast and is not related to the field gradient. Hence, it can be applied to both MRIL and CMR. On the other hand, the variable TE method requires a well-defined field gradient, and is often limited to tools such as MRIL. However, it is possible to implement the variable TE method for tools with complex field gradients. Elaborate schemes have been developed for CMR and tested with reasonable success.

Flaum et al. [82] suggested methods to determine the gas saturation, using NMR signal amplitude, echo difference, and echo ratio (variable TE). Freedman et al. [89] later proposed a formula using the density porosity and the NMR porosity to compute the gas saturation. Flaum et al.'s method [84] was recently extended, using variable TE, to estimate the oil saturation with reasonably good results [84].

Gas saturation

Flaum et al.'s NMR signal amplitude method [82] for the gas saturation determination is quite similar to Freedman et al.'s approach [89]. The latter provides a more extensive formulation, which we shall discuss as follows.

This method requires a good knowledge of the grain density and a good quality density porosity. Combining these with a knowledge of the hydrogen index and the T_1 value for gas, the gas saturation can be accurately determined.

Recall that the bulk density ρ_b is given by

$$\rho_b = (1 - \phi_{den})\rho_m + \rho_f\phi_{den}, \tag{5.17}$$

where ρ_m and ρ_f are the matrix and fluid densities, respectively, and the density porosity ϕ_{den} is

$$\phi_{den} = \frac{\rho_m - \rho_b}{\rho_m - \rho_f}. \tag{5.18}$$

When there is gas in the formation, Eq. (5.17) becomes

$$\rho_b = (1 - \phi)\rho_m + \rho_f(1 - S_g)\phi + \rho_g S_g \phi, \tag{5.19}$$

where S_g is the gas saturation in fractions, ρ_g is the gas density, and ϕ is the true total porosity. The ϕ_{NMR} can be written as follows:

$$\phi_{NMR} = \phi(1 - S_g) + S_g\phi \cdot HI \cdot P, \tag{5.20}$$

where HI is the hydrogen index, and P is the polarization factor

$$P = 1 - e^{-RD/T_1}, \tag{5.21}$$

where RD is the repeat delay time. Combining Eqs. (5.17), (5.19), and (5.20) to eliminate ϕ, we finally get

$$S_g = \frac{\phi_{den} - \phi_{NMR}}{\phi_{den}(1 - HI \cdot P) + \lambda \cdot \phi_{NMR}} \tag{5.22}$$

for the gas saturation, S_g, and

$$\phi = \frac{\phi_{den}(1 - HI \cdot P) + \lambda \cdot \phi_{NMR}}{(1 - HI \cdot P) + \lambda} \tag{5.23}$$

for the true total porosity, ϕ; where $\lambda = (\rho_f - \rho_g)/(\rho_m - \rho_f)$.

Freedman et al.'s method is not limited to tools with complex field gradients, and hence is applicable to MRIL as well. The approach of echo ratio proposed by Flaum et al. [82] for gas zone identification is specifically for tools with complex field gradients such as the CMR. It involves mapping out the field gradient distribution for application to the echo ratio analysis. Since it was further extended to solve for oil saturation, we shall combine the discussion in the following section.

Oil saturation

Flaum et al.'s method for gas/oil saturation determination [82, 84] utilizes the variable TE approach in a complex field gradient. To proceed, we should first understand how the T_2 echo train is affected in a complex field gradient. We know that for a constant field gradient, G, the amplitudes of a spin echo train can be written as:

$$A_i(t_i, \tau, D) = \sum_{j=1}^{m} f_j e^{-t_i/T_j} e^{-t_i/T_{2D}}, \tag{5.24}$$

where A_i is the amplitude of the i-th echo, f_j is the partial porosity whose spins have a T_2 relaxation time of T_j ($j = 1, ..., m$, and there are m relaxation times), and T_{2D} is the constant gradient term as defined previously (Eq. (5.16)):

$$T_{2D} = 3(\gamma^2 G^2 D \tau^2)^{-1}. \tag{5.25}$$

If we make the assumption that the sample can be subdivided into small volume elements, each of which has a more or less constant field gradient, and that all these contributions can be integrated, then we have the echo amplitude as

$$A_i(t_i, \tau, D) = \int J(G) \sum_{j=1}^{m} f_j e^{-t_i/T_j} e^{-t_i/T_{2D}} dG \tag{5.26}$$

where $J(G)dG$ is the infinitesimal portion which has a gradient of G. Note this assumption is only an approximation. There are some intricate details about the diffusion of spins in a non-uniform gradient field, which we shall discuss in the last chapter.

Neglecting any susceptibility contrast induced internal field gradient, we can place the integration inside the summation:

$$A_i(t_i, \tau, D) = \sum_{j=1}^{m} f_j e^{-t_i/T_j} \int J(G) e^{-t_i/T_{2D}} dG \tag{5.27}$$

To obtain the field gradient distribution $J(G)$, we can use a bottle of water sample whose $T_{2B,w}$ is known accurately beforehand. Since this is bulk water, the summation of various T_j reduces to a single term, $e^{-t_i/T_{2B,w}}$. Dividing both sides of Eq. (5.27) by $e^{-t_i/T_{2B,w}}$, we get

$$\tilde{A}_i(t_i, \tau, D) = \frac{A_i}{e^{-t_i/T_{2B,w}}} = \int J(G) e^{-t_i/T_{2D,w}} dG. \tag{5.28}$$

Upon close examination of Eq. (5.28), it is easily realized that we now have a problem very similar to inverting an echo train to get a T_2 amplitude distribution. The "echo" amplitude is $\tilde{A}_i$, the "T_2" is $T_{2D,w}$ (which is a function of G as shown in Eq. (5.16)), and the "T_2 amplitude" is simply $J(G)$.

Once $J(G)$ is found, it is straightforward to establish the following function $F(t_i, \tau, D)$ for any fluid or gas with a different diffusion constant, D, and hence, T_{2D},

$$F(t_i, \tau, D) = \int J(G) e^{-t_i/T_{2D}} dG. \tag{5.29}$$

To determine the gas zone, Flaum et al. [82] tried to extract information of the apparent diffusion constant, D, of the formation fluid from the NMR logs. As the diffusion constant becomes significantly higher than that of the water, it is taken as a sign of entering the gas zone.

The echo ratio method was used to determine the apparent D. The ratio of echo amplitudes taken at the same elapsed time t for two different echo trains measured with a short and a long τ is given as follows:

$$\frac{A_i(t_i, \tau_{short}, D)}{A_i(t_i, \tau_{long}, D)} = \frac{F(t_i, \tau_{short}, D)}{F(t_i, \tau_{long}, D)}, \tag{5.30}$$

where the summation part before the integral in the RHS of Eq. (5.27) has been canceled.

Flaum et al. [82] found empirically that the apparent D is given by

$$D = aM^2 + bM, \tag{5.31}$$

where M is the first moment of the echo amplitude ratio distribution minus 1:

$$M = \sum_i t_i [F(t_i, \tau_{short}, D)/F(t_i, \tau_{long}, D) - 1]. \tag{5.32}$$

The constants a and b can be determined beforehand with the bottle of water experiment. The ratio of echo amplitudes was calculated using the fitted decay curves to avoid the noise problem. Flaum et al. [82] found that the method provided a good prediction of the gas zone, although the extracted apparent D for the gas was about a factor 3 less than it should be in downhole conditions. Considering the restricted diffusion effect and the presence of water and oil, such a reduction is not unexpected.

To determine the oil saturation, Flaum et al. [84] suggested a combination of time domain analysis and the variable TE method, which they called *Enhanced Diffusion Analysis* (EDA).

The difference of two echo trains with different TE is obtained as follows:

$$\begin{aligned} \Delta A_i &= A_i(t_i, \tau_{short}, D) - A_i(t_i, \tau_{long}, D) \\ &= \sum_{j=1}^{m} f_j e^{-t_i/T_j}[F(t_i, \tau_{short}, D_w) - F(t_i, \tau_{long}, D_w)] \\ &+ f_{oil}(1 - e^{-RD/T_{1,oil}})e^{-t_i/T_{2,oil}}[F(t_i, \tau_{short}, D_{oil}) - F(t_i, \tau_{long}, D_{oil})] \end{aligned} \tag{5.33}$$

where RD is the repeat delay time used in the acquisition of the echo trains.

The actual procedure for carrying out the *Enhanced Diffusion Analysis* is somewhat involved. Eq. (5.33) is first solved in the water zone for the T_2 signal of water by assuming a known D_w. This is accomplished through trial and error by matching the echo difference to the predicted water signal. Then the same shape of the T_2 distribution for water is assumed in the oil zone to solve for the oil saturation using Eq. (5.33). As some of the water residing in the small pores does not contribute to the water signal in the difference of the two echo trains due to restricted diffusion, proper steps should be taken to take this into consideration. Flaum et al. [84] have shown that this method can be successfully carried out in a finite interval where the rock fabrics are similar between the water and oil zones.

Thus, we see that through an elaborate and complicated scheme, the variable TE method can also be implemented for a tool with complex field gradients. The physics involved, however, is not as straightforward as that for a constant field gradient.

5.6 Oil viscosity estimation

The NMR log is the only log so far that provides information on the oil viscosity. As we discussed in Chapter 4, the T_1 or T_2 of oil is related to viscosity, η, through the following relationship (Vinegar [220]):

$$T_{1,2} = \frac{C_1 T_K}{298\eta}, \tag{5.34}$$

where $T_{1,2}$ is in seconds, and usually refers to the logarithmic average of $T_{1,2}$; T_K, the absolute temperature, is in degrees Kelvin; η, the viscosity, is in centipoise (cp); and C_1 is an empirical fitting constant whose value varies from 2.8 to 1.2 depending upon whether the oil sample is oxygen free or not (see discussion in Chapter 3).

The diffusion constant of the oil is also related to the viscosity (Vinegar [220]):

$$D = \frac{C_2 T_K}{298\eta} \tag{5.35}$$

where D is in units of 10^{-5}cm^2/sec, T_K and η are in the same units as described previously, and C_2 is approximately 1.4 for oil and 2 for water, respectively.

As a first order approximation, if one can identify the T_2 peak related to the oil signal, one can simply use Eq. (5.34) directly to estimate the oil viscosity. However, after some thought, one should realize that Eq. (5.34) is obtained in the laboratory with a homogeneous magnetic field, whereas the apparent T_2 peak obtained from the NMR logs is most likely affected by the magnetic field gradient G as described in the following:

$$\begin{aligned} \frac{1}{T_2^\dagger} &= \frac{1}{T_{2,int}} + \frac{1}{T_{2D}} \\ &= \frac{1}{T_{2,int}} + \frac{1}{3}(\gamma^2 G^2 D \tau^2) \end{aligned} \tag{5.36}$$

Thus for large viscosity, or small diffusion constant, and when the contribution from the diffusion term $1/T_{2D}$ is small, the first order approximation of ignoring the diffusion contribution and using Eq. (5.34) directly would be acceptable. If this is not satisfied, then more elaborate analysis is necessary.

5.6.1 Multiple TEs and repeat delay times

Using a combination of multiple TE and dual wait time scheme, Chen et al. [58, 59] suggested a method which provides quite successful estimates of oil viscosity from the NMR logs. Their procedure consists of the following three steps:

1. Separate the oil signal from the water signal using time domain analysis of a dual wait time run.

2. Determine the apparent T_2 for oil for different TE runs, assuming it is the non-wetting phase.

3. Estimate the oil viscosity using $T_{2,int}$ or T_{2D}.

The first step is accomplished by conducting three different TE (2τ) passes, each with a dual wait time, i.e., a long and a short wait time. The echo train of the short wait time is subtracted from that of the long wait time in the time domain. The difference of the echo trains is assumed to be primarily coming from the oil signal for the oil bearing zone. It can be inverted to get a T_2 distribution, primarily for oil with a unimodal shape, or it can be fitted with a single exponential. To ensure the residual water signal in the difference of the echo trains is removed, the data of the first 24 ms is omitted in the processing. The apparent T_2 is then determined for each different TE. This completes the second step. The last step is to plot the reciprocal of this apparent T_2, or apparent relaxation rate, as a function of τ^2:

$$\begin{aligned}\frac{1}{T_2^\dagger} &= \frac{1}{T_{2,int}} + \frac{1}{3}(\gamma^2 G^2 D)\tau^2 \\ &= A + B\tau^2 \qquad (5.37)\end{aligned}$$

where the intercept A, determined from the linear regression fit, gives $1/T_{2,int}$. For the non-wetting phase, it is the same as the bulk value, and hence, Eq. (5.34) can be used to estimate the viscosity. The slope B, of such a fit, gives a diffusion constant D, provided the gradient G is known precisely. From the value of D and Eq. (5.35), the oil viscosity can also be evaluated. Thus we have two independent methods to estimate oil viscosity, i.e., one from the intercept and Eq. (5.34), and the other from the slope and Eq. (5.35). These two estimates can be used as a cross check of the procedure.

When there are strong internal field gradients due to the presence of ferro- and paramagnetic materials, the value of G may be significantly modified. In this case, the oil viscosity derived from the intercept is more reliable.

The oil viscosity η can also be obtained directly from a nonlinear fit of the following equation (derived by combining Eqs. (5.34), (5.35), and (5.37)):

$$\frac{1}{T_2^\dagger} = A_1 \cdot \eta + A_2 \tau^2 \cdot \eta^{-1} \qquad (5.38)$$

where $A_1 = 298/(C_1 T_K)$ and $A_2 = \gamma^2 G^2 C_2 T_K/(298 \cdot 3)$. Naturally, in this approach, you have to assume that G is accurately known and the internal gradient is insignificant.

5.6.2 Summation of echoes

The summation of echoes was first suggested by Sezginer et al. [199] as a permeability indicator. Later work by Chen et al. [60] showed that the summation of echoes is a very useful petrophysical parameter which can be used for estimating the T_1 of oil, hence for oil viscosity determination, as well as depth-matching, log quality control, etc.

To understand what kind of information is contained in the summation of echoes, let us first examine this quantity analytically. The echo amplitude, g_i, for a T_2 CPMG measurement is typically expressed as:

$$g_i = \sum_{j=1}^{m} f_j e^{-it_E/T_j} + \epsilon_i \tag{5.39}$$

where f_j is the partial porosity for T_2 relaxation time $T_{2,j}$ (There are m relaxation times, $j = 1, ..., m$, and $\sum_j f_j = \phi$, where ϕ is the total porosity), t_E is TE, and ϵ_i is the noise for the i-th echo amplitude. Thus, if we denote the summation of echoes by SE, we get,

$$\begin{aligned} SE &= \sum_{i=1}^{n}\sum_{j=1}^{m} f_j e^{-it_E/T_{2,j}} + \sum_{i=1}^{n} \epsilon_i \\ &= \sum_{j=1}^{m} f_j \frac{1 - e^{-nt_E/T_{2,j}}}{e^{t_E/T_{2,j}} - 1} + \sum_{i=1}^{n} \epsilon_i \\ &\approx \sum_{j=1}^{m} f_j \frac{T_{2,j}}{t_E} + \mathcal{O}[\sqrt{n}\sigma] \\ &= \frac{\phi\langle T_2\rangle}{t_E} + \mathcal{O}[\sqrt{n}\sigma] \end{aligned} \tag{5.40}$$

where σ is the statistical average of the noise, and we have made the following assumptions

$$t_E \ll T_{2,j} \quad \longrightarrow \quad e^{t_E/T_{2,j}} \approx 1 + \frac{t_E}{T_{2,j}} \tag{5.41}$$

and

$$nt_E \gg T_{2,j} \quad \longrightarrow \quad e^{-nt_E/T_{2,j}} \approx 0. \tag{5.42}$$

Thus we can immediately see the benefit of this method that the summation of echoes is proportional to the porosity times the average of the T_2. The simple average of T_2 is weighted heavily toward the long relaxation times. This is the reason why the summation of echoes can be used as a permeability indicator. However, there are situations when $\langle T_2 \rangle$ does not correlate well with permeability.

To apply the summation of echoes to estimate the T_1 of the oil, two NMR logs with different wait times should be performed. We note that in the water and oil zone, the echo amplitudes for long and short wait times, $g_{i,L}$ and $g_{i,S}$, are given respectively as:

$$\begin{aligned} g_{i,L} &= \sum_{j=1}^{m} f_{j,w} e^{-it_E/T_{2,j}} + \phi_o H I_o \alpha_L e^{-it_E/T_{2,o}} \\ g_{i,S} &= \sum_{j=1}^{m} f_{j,w} e^{-it_E/T_{2,j}} + \phi_o H I_o \alpha_S e^{-it_E/T_{2,o}} \end{aligned} \tag{5.43}$$

where ϕ_o is the porosity associated with oil, HI_o is the hydrogen index for oil, and $\alpha_L = 1 - e^{-RD_L/T_{1,o}}$ and $\alpha_S = 1 - e^{-RD_S/T_{1,o}}$ are polarization factors for long and short wait times for oil, respectively. Thus the corresponding summations of echoes for long and short wait times are:

$$\begin{aligned} SE_L &\approx \frac{\phi_w \langle T_{2,w} \rangle}{t_E} + \frac{HI_o \alpha_L \phi_o T_{2,o}}{t_E}(1 - e^{-nt_E/T_{2,o}}) \\ SE_S &\approx \frac{\phi_w \langle T_{2,w} \rangle}{t_E} + \frac{HI_o \alpha_S \phi_o T_{2,o}}{t_E}(1 - e^{-nt_E/T_{2,o}}) \end{aligned} \tag{5.44}$$

If we assume that the first term for the water, i.e.,

$$f_w = \frac{\phi_w \langle T_{2,w} \rangle}{t_E} \tag{5.45}$$

is known from the T_2 distributions obtained from the regular inversion, then the ratio

$$R = \frac{SE_L - f_w}{SE_S - f_w} = \frac{\alpha_L}{\alpha_S} = \frac{1 - e^{-RD_L/T_{1,o}}}{1 - e^{-RD_L/T_{1,o}}} \tag{5.46}$$

can be used to estimate T_1 of the oil.

This method is somewhat better than the regular time domain analysis. The latter requires prior knowledge of the tool field gradient to relate $T_{1,o}$ and $T_{2,o}$, and needs to solve for ϕ_o and $T_{1,o}$ simultaneously. The process is also strongly affected by the noise level and the different methods of regularization.

The summation of echoes method, on the other hand, needs to solve for only one parameter, T_1, thus providing a good estimate of oil viscosity. The quantities, SE_L and SE_S have been determined with very high precision, resulting from the effective reduction of noise in the time domain. Their difference is enhanced by the factor $T_{2,o}(1 - e^{-nt_E/T_{2,o}})/t_E$. Furthermore, the inaccuracy in the quantity for water, f_w, is relatively unimportant, because $\langle T_{2,w}\rangle \ll T_{2,o}$.

Because of the high precision of SE, it can be used for depth matching, especially for passes with high noise level where depth matching is difficult. Note in Eq. (5.44), because $\langle T_{2,w}\rangle \ll T_{2,o}$, the value of SE is quite insensitive to the first term, but very sensitive to the second term. Thus it can also be used for log quality control to determine whether the wait time is sufficient in the logging passes.

5.6.3 Constituent viscosity model

Freedman et al. [90] recently suggested a pore fluids characterization method using a suite of spin-echo measurements with different echo spacings, polarization times, applied field gradients, and number of echoes to invert the T_2 distributions of various pore fluids. They called this, *magnetic resonance fluid* (MRF) characterization method.

Freedman et al.'s approach is much more ambitious than Chen et al.'s method. Not only does it yield information on the oil viscosity, it also yields the T_2 distribution of oil, hence the oil saturation, as well as the gas saturation, if there is any. It uses six or more measurements for one sample or one formation interval to solve for all unknowns in the amplitudes of the T_2 distributions for water, oil, and gas (plus oil-based mud filtrate, if present). Thus, unlike Chen et al.'s approach where only a single T_2 is used for oil, Freedman et al.'s method provides a T_2 distribution for oil. This makes a significant difference when the T_2 distribution for oil is multi-modal rather than unimodal which can be represented by a single T_2.

Freedman et al.'s method has not been widely tested yet. The ability of their method to invert the relative amplitudes for overlapping water and oil signals hinges upon the important hypothesis of the *Constituent Viscosity Model* (CVM). This CVM hypothesis links the constituents of the $T_{1,2}$ relaxation times of a crude oil with those of the diffusion coefficients of the same crude oil, as described in the following.

If the macroscopic empirical correlation of T_2 (or more specifically $T_{2,LM}$, the logarithmic average of T_2) is given by the following equation:

$$T_{2,LM} = \frac{aT_K}{\eta} \tag{5.47}$$

where $T_{2,LM}$ is the logarithmic average over $T_{2,j}$:

$$\log T_{2,LM} \equiv \sum_{j=1}^{m} f_j \log T_{2,j}, \qquad \sum_{j=1}^{m} f_j = 1, \tag{5.48}$$

then the *Constituent Viscosity Model* defines a constituent viscosity, η_j, given by the following relationship,

$$T_{2,j} = \frac{aT_K}{\eta_j} \tag{5.49}$$

with the natural consequence:

$$\log \eta_{LM} \equiv \sum_{j=1}^{m} f_j \log \eta_j = \log \eta. \tag{5.50}$$

Note that the constituent viscosity, η_j, is strictly and operationally defined by Eq. (5.49), and has no real physical meaning except to indicate some kind of smeared out distribution of the components of a mixture. For example, a crude oil could be a mixture of three components, while its T_2 distribution could have many $T_{2,j}$ relaxation times, and thus as many η_j. There could be as many as, say, 20 η_j, whereas there are only three true physical components in the mixture.

The critical hypothesis that CVM makes is the following: In a Stejskal and Tanner type pulsed field gradient spin-echo experiment (PFGSE) [205, 209], the constituent diffusion coefficients, D_j, would have exactly the same amplitude, f_j, from the inversion of the following signal ratios obtained from the PFGSE experiments:

$$\frac{M}{M_0} = \sum_{j=1}^{m} f_j e^{-\gamma\delta^2 G^2(\Delta - \frac{1}{3}\delta)D_j} \tag{5.51}$$

as those in Eq. (5.48), if the D_j is chosen in the following manner

$$D_j = \frac{bT_K}{\eta_j}. \tag{5.52}$$

With this hypothesis, it easily follows that

$$\eta = \frac{aT_K}{T_{2,LM}} = \frac{bT_K}{D_{LM}} \equiv \eta_{LM} \tag{5.53}$$

where

$$\log D_{LM} \equiv \sum_{j=1}^{m} f_j \log D_j, \tag{5.54}$$

and that

$$\frac{D_j}{T_{2,j}} = \frac{b}{a} = \frac{D_{LM}}{T_{2,LM}}. \tag{5.55}$$

Thus an important consequence of this CVM hypothesis is that the T_2 and D distributions are not independent; i.e., if one is known, the other one can be calculated. The CVM was mainly based on the experimental work of Lo et al. [144]. There is some experimental evidence to support the CVM hypothesis which links the T_2 and D distributions, but it has not been extensively tested.

With the CVM hypothesis, Freedman et al. were able to successfully invert the T_2 amplitudes of water, oil, gas, and oil-based mud filtrate (OBMF) using the following multifluid relaxational model:

$$\begin{aligned} g_i^p &= \sum_{j=1}^{m} f_{w,j} e^{-it_{E_p}/T^{\dagger}_{2w,j}(p)} (1 - e^{-RD_p/\xi T_{2,j}}) \\ &+ \sum_{j=1}^{m} f_{o,j} e^{-it_{E_p}/T^{\dagger}_{2o,j}(p)} (1 - e^{-RD_p/T_{1,o}}) \\ &+ f_g e^{-it_{E_p}/T^{\dagger}_{2g}(p)} (1 - e^{-RD_p/T_{1,g}}) \\ &+ f_{OBMF} e^{-it_{E_p}/T^{\dagger}_{2,OBMF}(p)} (1 - e^{-RD_p/T_{1,OBMF}}) \end{aligned} \tag{5.56}$$

where g_i^p is the i-th echo of the p-th experiment, t_{E_p} the echo spacing, RD_p the repeat delay time, ξ the T_1/T_2 ratio for water, and $f_{w,j}$, $f_{o,j}$, f_g and f_{OBMF} the partial amplitudes for water, oil, gas, and oil-based mud filtrate (OBMF), respectively. The apparent T_2's for water ($T^{\dagger}_{2w,j}(p)$), oil ($T^{\dagger}_{2o,j}(p)$), gas ($T^{\dagger}_{2g}(p)$), and OBMF ($T^{\dagger}_{2,OBMF}(p)$) are given as follows:

$$\frac{1}{T^{\dagger}_{2w,j}(p)} = \frac{1}{T_{2w,j}} + \frac{1}{12}\gamma^2 G_p^2 t_{E_p}^2 D_w \tag{5.57}$$

$$\frac{1}{T^{\dagger}_{2o,j}(p)} = \frac{1}{T_{2o,j}} + \frac{1}{12}\gamma^2 G_p^2 t_{E_p}^2 D_o \tag{5.58}$$

$$\frac{1}{T^{\dagger}_{2g}(p)} = \frac{1}{T_{1,g}} + \frac{1}{12}\gamma^2 G_p^2 t_{E_p}^2 D_g \tag{5.59}$$

$$\frac{1}{T^{\dagger}_{2,OBMF}(p)} = \frac{1}{T_{2,OBMF}} + \frac{1}{12}\gamma^2 G_p^2 t_{E_p}^2 D_{OBMF} \tag{5.60}$$

where aside from the diffusion coefficient for gas and water, the diffusion coefficients for oil and OBMF are related to their corresponding T_2 relaxation times through the CVM equations.

Furthermore, the constituent viscosity model can be extended to include the live oils by incorporating the empirical correction for gas oil ratio discussed in Chapter 3.

5.7 Logging guidelines

Because of the complexity of NMR logging and the large number of tool parameters, it is essential to have a pre-logging plan to ensure all logging parameters are set properly to obtain the desired information from the logging run. Most NMR logging companies have a pre-logging planning kit which allows the users to calculate the proper logging parameters for the logging run. Here, we shall mention some of the general guidelines prior to the logging job. For those logging parameters which are tool specific, the readers should refer to the instruction manuals supplied by the logging companies.

Knowing the tool characteristics

It is very important to know the tool characteristics, its capabilities, and its limitations, and to determine if the logging objectives can be accomplished using the specific tool. For example, some tools (e.g. Halliburton/Numar's MRIL) are designed as centralized tools, whereas others (e.g. Schlumberger's CMR) are pad-type tools. For the centralized tools, high salinity of the borehole fluid decreases the efficiency of the tool (lowered Quality Factor, Q). The reduced Q of the transmitting/receiving coil results in decreased signal to noise (S/N) ratio. For certain hole sizes, a mud excluder can be slipped over the tool to minimize this problem. In larger holes, it may be necessary to use a fresh water slug for proper operation of the tool.

The pad type tool does not have the borehole salinity problem. However, depending on the depth of investigation, the tool may be probing a shallow depth into the formation, thus measuring the invaded mud filtrate rather than the formation fluid.

Different tools have very different magnetic field distributions. Depending on these magnetic field distributions, some applications of NMR logging discussed previously may be possible, whereas others may not be applicable.

Tool calibrations and preparations

Prior to logging, it is essential to make sure the tool is properly calibrated. Improper calibration can affect the porosity reading. Other preparation procedures are often recommended by the service companies. They need to be closely followed to ensure the quality of the data.

Establishing the proper repeat delay time

The repeat delay time is designated as RD, or sometimes as TR (Time to Repeat), or WT (Wait Time). This is the time between the successive pulse excitations. To ensure that all the spins are fully recovered from the previous excitation and polarized again along the direction of the applied magnetic field, it is essential to have a long enough RD. The value of RD is dictated by the longest T_1 value of the formation. A good rule of thumb is to choose RD to be three times the longest T_1 of the formation fluids. If too short a repeat delay time is used, porosity and free fluid index will be underestimated. If RD is too long, logging speed will be unnecessarily reduced. Some of the modern NMR logging tools have a long magnet that prepolarizes the spins in the formation, thus increases the logging speed.

All the NMR logging service companies have the means to determine the proper RD or wait time from measurements made in the well. These measurements have been adequate in most wells where water based muds were used, and moderate well bore temperatures ($< 200°$F) were encountered. For higher temperatures and/or oil based muds, it may be necessary to have prior laboratory or field site measurements on reservoir fluids to determine the T_1 and T_2 in order to ensure that the proper repeat delay times are used.

Well site measurements are recommended where a good filtrate sample can be obtained usually from the mud company who uses a filtrate press. Where a good filtrate sample cannot be obtained, mud and produced fluid samples should be kept and measured later in the laboratory. From these measurements, corrections can later be made to the logs if too short a repeat delay time was used.

Getting a filtrate sample from an oil based mud is difficult but should be available from the mud companies. If a good filtrate sample cannot be obtained from the mud company, it is recommended to centrifuge the samples and draw off the top (less dense) fraction.

Where the surface temperatures are significantly different from those in the wellbore, it is recommended to slowly lower the tool into the hole so as to allow sufficient time for temperature stabilization.

Stationary measurements

Stationary measurements should be performed in the well to obtain an estimate for T_1 and to improve the default setting of the T_1/T_2 ratio. This parameter is often set at 1.5. (NOTE: In some cases, the T_1/T_2 ratio can be much higher than 1.5. This is not a concern if the wait time is sufficient for the logging conditions.) Stationary measurements should be recorded long enough for the signal/noise to reach at least 20.

Log quality controls

The logging engineer (unless told otherwise) will choose a logging speed to minimize acquisition time while maintaining an acceptable resolution and signal to noise ratio. The highest resolution and signal to noise ratio are achieved at the slowest speeds. Consequently, it is recommended (where hole conditions permit) to run the tool as slowly as possible. The data quality improvement can be tested during tool setup by choosing a short interval and running the tool first at the speed recommended by the logging company, and then at a half or a quarter of that speed.

Tool repeatability

The measurement is statistical in nature and does not repeat exactly. When the borehole is not washed out or rugose, porosity should repeat within 1 p.u. and FFI should repeat within 0.5 p.u.

Log data consistency checks

NMR porosity should be comparable to matrix corrected neutron/density porosity in clean zones where there is no heavy oil and gas. Shale and heavy oil will cause a reduction in NMR porosity as compared to crossplot porosity.

Chapter 6

NMR data acquisition and inversion

Processing of NMR data involves detailed knowledge of the data acquisition procedure, how different pulsing sequences are averaged, and the proper phase rotation to obtain the echo train signal. There are also various schemes for data inversion to obtain the T_2 distributions which deserve a careful review.

In the following, we shall describe the general procedure of acquiring NMR data for T_2 CPMG type measurement in the laboratory, and follow with a discussion on various data inversion methods. Different NMR logging tools have slightly different pulsing sequences. As a result, the way the signals are averaged may affect the data storage format. The readers can refer to vendor's manuals for details. However, the basic general principles of CPMG data acquisition are the same.

6.1 Data acquisition

In addition to the detection of the FID and the CPMG spin echo signals, we need to digitize, store, and process these signals. For a typical CPMG spin echo train, there are frequently hundreds or even thousands of echoes. It is not possible, and not necessary either, to digitize the entire echo profile, only the peaks of the spin echo signal need to be measured and stored. There are two ways to record the signal: the single channel and the quadrature detection methods.

6.1.1 Signal detection

In the single detection method, the rf signal from the sample is fed into a single phase-sensitive detector for determining the frequency difference between the sample signal and the carrier frequency of the rf pulse (i.e., the mean Larmor frequency which we set for the measurements). Since a single phase-sensitive detector can determine only the magnitude and not the sign of this difference, we won't be able to determine whether the offset frequency is positive or negative. Hence, a single spectral feature, after Fourier transform, can be detected at two different frequencies symmetrically located about the carrier frequency ν_0. The process of reflecting the feature at, say, $\nu_0 + \delta\nu$ to $\nu_0 - \delta\nu$, is called folding.

To avoid confusion, the offset of the carrier frequency must be chosen at one end of the spectral region. Signals which occur outside the range of the measurement become "folded" and appear as mirror signals, usually of indeterminate phase.

Because of its inability to determine the phase differences, single detection is not used in NMR logging. Rather, quadrature detection is the method normally used for signal detection.

Using quadrature detection, the incoming sample signal is fed to two identical phase-sensitive detectors whose reference signals differ by exactly 90°. The resultant audio signals are amplified, passed through identical low-pass filters, digitized by a multiplexed analog to digital converter and stored in separate data memory blocks. Quadrature Fourier transformation produces a real and an imaginary spectrum in a manner analogous to single detection with the exception that now plus and minus frequencies relative to the rf carrier frequency can be distinguished.

Suppose the incoming sample signal, after mixing with the local oscillator frequency ω_c, has the form $\cos[(\omega_0 - \omega_c + \Delta\omega_B)t - \phi]$, where ω_0 is the mean Larmor frequency, ω_c is the carrier frequency, and $\Delta\omega_B$ is the bandwidth for all resonant spins. When $\omega_0 = \omega_c$, the mixed signal has the following form:

$$R(t) = \sum_i A_i(t)\cos(\omega_i t - \phi) \tag{6.1}$$

where the summation is over all frequencies ω_i within the resonance bandwidth from $-\Delta\omega_B/2$ to $+\Delta\omega_B/2$, and $A_i(t)$ is the amplitude for ω_i. When this is fed to the two phase-sensitive detectors, there would be a cos (real) channel as Eq. (6.1), and a sin (imaginary) channel as

$$I(t) = \sum_i A_i(t) \sin(\omega_i t - \phi) \tag{6.2}$$

In the case of a FID, $A_i(t)$ will be an exponential decay function. The rapid decay of the FID is the result of the summation of various frequencies within the resonance band. In the case of CPMG measurements, the sample signal will have the appearance as shown in Figure 6.1. Notice that (a) and (b) of Figure 6.1, representing the output of the two phase-sensitive detectors, have exactly the same envelop $\sum_i A_i(t)$. This is a case where the phase angle ϕ is exactly zero. Hence, if we digitize the output curve at a time interval of 2τ, i.e., where the peak of the spin echo is, Figure 6.1(a) will be $\sum_i A_i(t) \cos 0°$, i.e., tracing out the profile of $\sum_i A_i(t)$, and Figure 6.1(b) will be $\sum_i A_i(t) \sin 0°$, tracing out the dc base line.

In the T_2 CPMG data storing process, only the signal within a very small time window at the peak of the spin echo is digitized for storage. Hence, Figures 6.1(a) and (b) will be further simplified to Figures 6.2(a) and (b). Figure 6.2(a) shows the T_2 echo train signal without any noise, and it is on this one that we perform our NMR inversion computation to obtain the T_2 amplitude distribution. Figure 6.2(b) simply shows the dc base line. Of course, in real log data, both of them will be masked by noise fluctuations.

If the phase angle ϕ is not zero, then the two channels will have the appearance as shown in Figure 6.3, and after sampling the signal every 2τ, they become Figure 6.4, which shows the two channels of data normally acquired during the NMR logging run.

The T_2 CPMG measurements of the logging runs are always collected in pairs, known as Phase-Alternated Pairs (PAPs), in order to cancel the ringing and baseline offset. The first echo train is acquired with the normal settings, whereas in the second set the phase of the transmitter pulse is changed 180° to give spin echoes of negative amplitude. The two sets of CPMGs are referred to as the plus phase and minus phase. These pairs are combined to give PAPs which eliminate the effects of ringing and baseline offset. In the laboratory where measurement time is not an issue, more refined measurements are possible and signals are almost always collected in four quadrants, instead of just two, to completely eliminate phase offset errors.

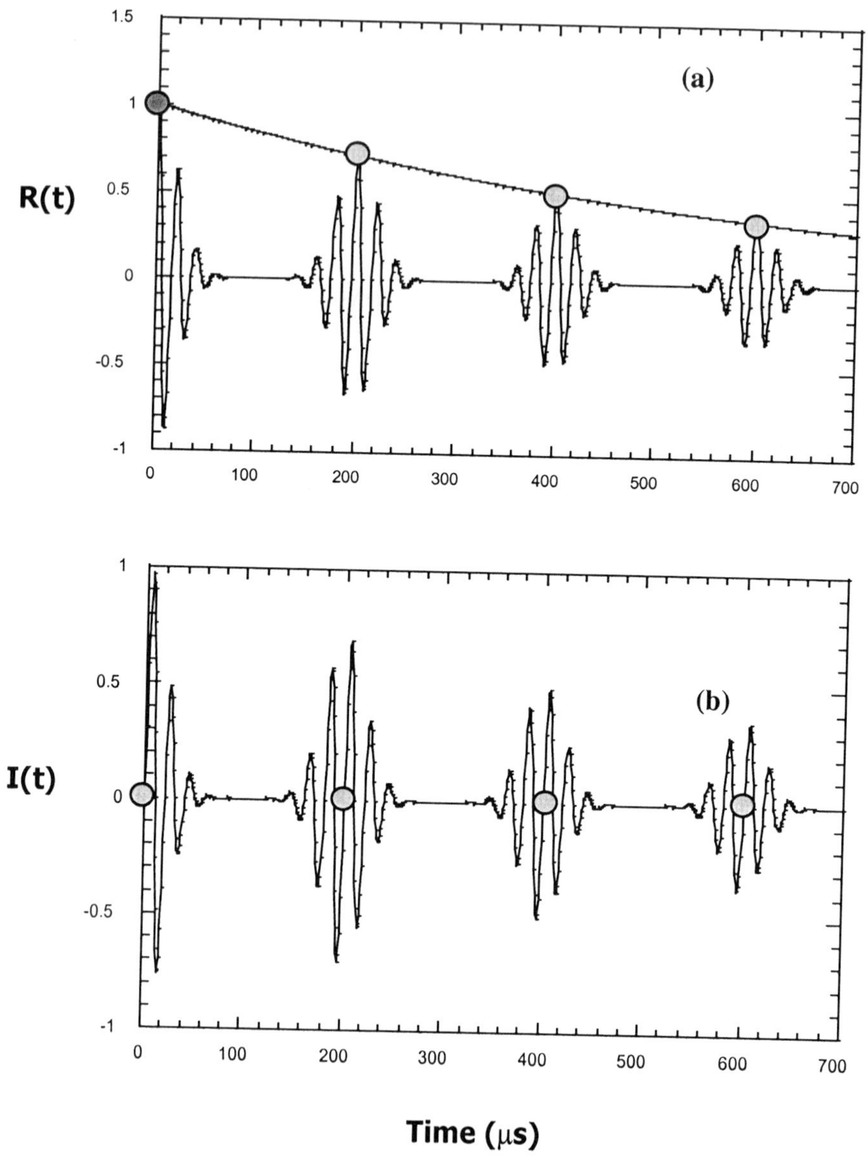

Figure 6.1: The two quadrature channels (a) cos (real) and (b) sin (imaginary) when the phase angle ϕ is zero

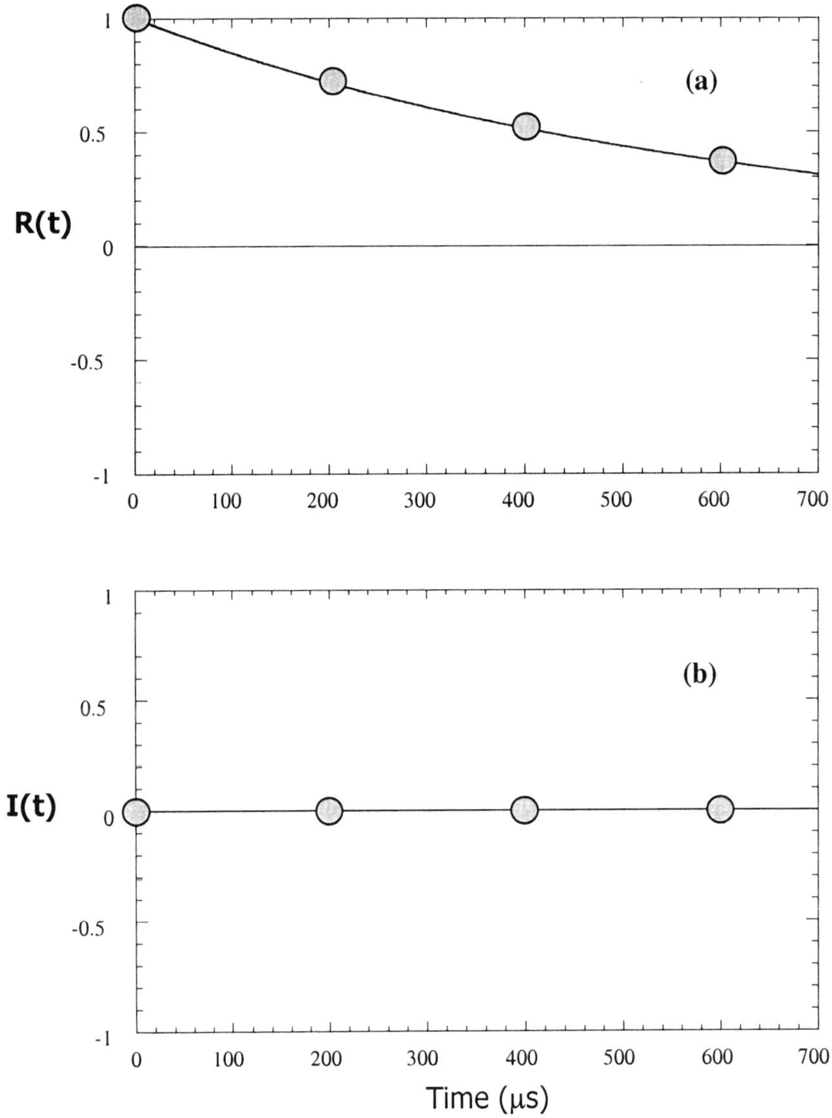

Figure 6.2: The two quadrature channels of Figure 6.1 after collecting the signal at the peak of the spin echo.

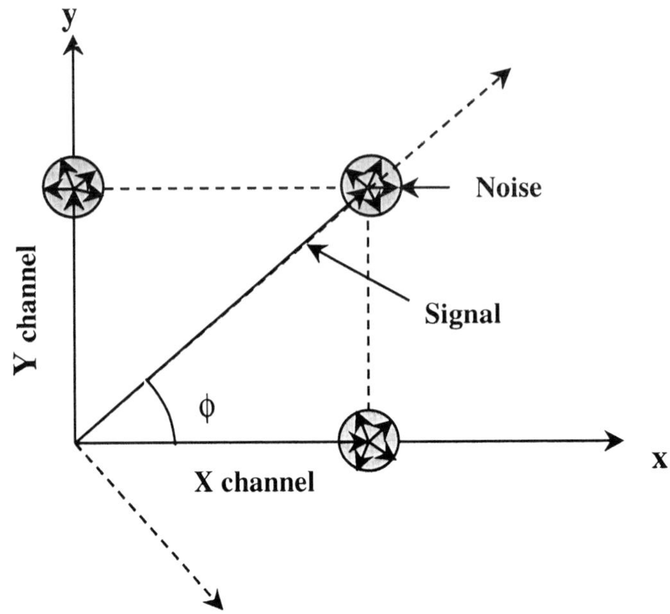

Figure 6.5: The echo signal magnitude and noise in polar coordinates and their relationship to X and Y channels.

6.1.2 Phase rotation

The two channels of recorded data are typically referred to as X and Y channels. To compute the phase angle (see for example, Freedman [88]), we do the following,

$$X_j = S_j \cos\phi + \varepsilon_j^X \tag{6.3}$$

$$Y_j = S_j \sin\phi + \varepsilon_j^Y \tag{6.4}$$

where X_j and Y_j are the amplitudes of the j-th echo in the X and Y channels, respectively, S_j the true magnitude of the echo, ϕ the phase angle, and ε_j^X and ε_j^Y the noise for the j-th echo in X and Y channels, respectively. Figure 6.5 shows their relationship in polar coordinates. If we sum up all echoes (assuming n echoes) in the echo train, we get,

$$\sum_{j=1}^{n} X_j = \left(\sum_{j=1}^{n} S_j\right)\cos\phi + \left(\sum_{j=1}^{n} \varepsilon_j^X\right) \tag{6.5}$$

$$\sum_{j=1}^{n} Y_j = \left(\sum_{j=1}^{n} S_j\right)\sin\phi + \left(\sum_{j=1}^{n} \varepsilon_j^Y\right) \tag{6.6}$$

Because the noise is random with equally probable positive and negative values, the second term in Eqs. (6.5) and (6.6) eventually approaches zero when the averaged noise is computed. Taking the ratio of (6.6) to (6.5), we have,

$$\phi = \tan^{-1}\left(\sum_{j=1}^{n} Y_j \Big/ \sum_{j=1}^{n} X_j\right) \tag{6.7}$$

Once the phase angle is known, the coordinates can be rotated using the following equations to obtain the signal channel and the noise channel, as shown in Figure 6.6

$$X_j \cos\phi + Y_j \sin\phi = S_j + (\varepsilon_j^X \cos\phi + \varepsilon_j^Y \sin\phi) \tag{6.8}$$

$$-X_j \sin\phi + Y_j \cos\phi = (-\varepsilon_j^X \sin\phi + \varepsilon_j^Y \cos\phi) \tag{6.9}$$

where (6.8) represents rotation to the signal channel, (6.9) rotation to the noise channel, and the quantities within the parentheses in (6.8) and (6.9) are noise.

In most cases, the summation in Eqs. (6.5) and (6.6) is performed from the first through the last echoes to have good statistics in canceling the noise. This is especially true for large signals in which $\sum X_j$ (or $\sum Y_j$) is dominated by $\sum S_j \cos\phi$ (or $\sum S_j \sin\phi$). However, when we encounter shaly zones where porosities are small, the signal diminishes quickly after a few spin echoes and the rest of the echo train is just noise. The summation over noise, $\sum \varepsilon_j^X$ (or $\sum \varepsilon_j^Y$), actually has a standard deviation which increases as $\sqrt{n}$, even though the averaged noise, i.e., $\langle \varepsilon^X \rangle = (\sum \varepsilon_j^X)/n$ decreases as $1/\sqrt{n}$. Hence, in these situations, the computation of phase angle may not be accurate if all echoes are taken. The solution for this is to perform the summation up to the point where the echoes approach the baseline, then stop the summation and ignore the rest of the echoes which are essentially noise.

Sometimes, instead of X_j and Y_j, the echo magnitude $A_j = \sqrt{X_j^2 + Y_j^2}$ and the echo phase $\psi_j = \tan^{-1}(Y_j/X_j)$ for each individual echo are provided. Then, one has to compute X_j and Y_j from A_j and ψ_j and then perform the proper rotation to get the signal channel.

6.2 Data inversion

We have discussed previously in Chapter 4 that there are non-linear inversion schemes which use a two or three exponential model, and linear inversion

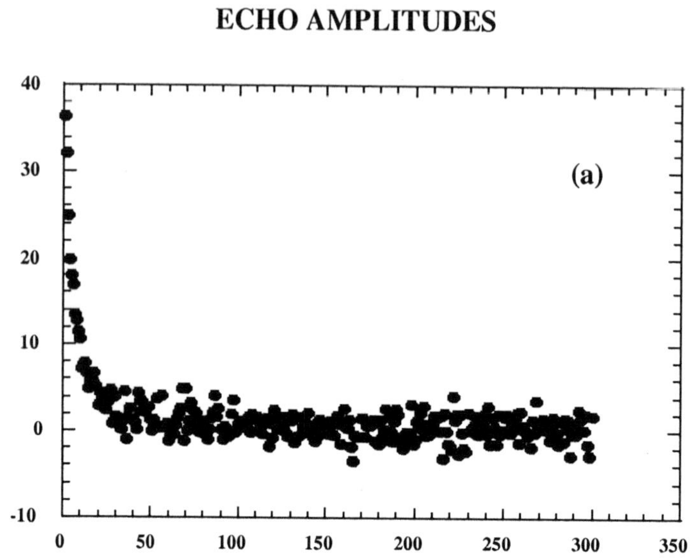

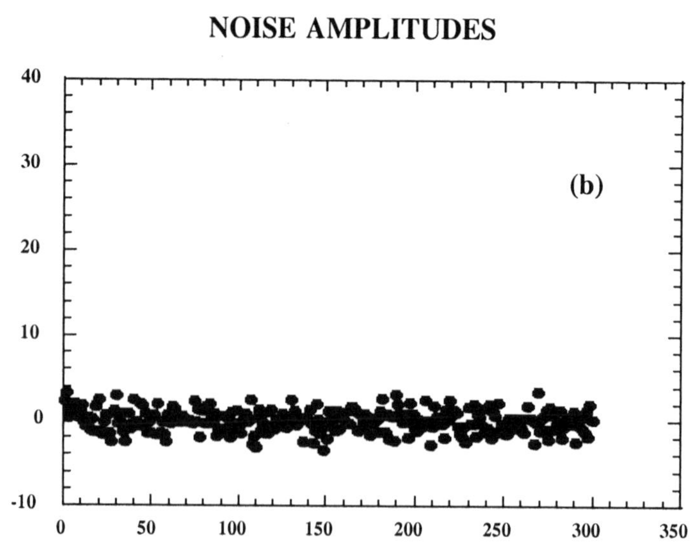

Figure 6.6: The signal and noise channels of a typical log measurement after the phase rotation.

schemes which assume a distribution of known T_2 relaxation times equally spaced on a logarithmic scale and solve for amplitudes. The CPMG T_2 echo trains obtained from NMR logging commonly use the linear inversion scheme to obtain T_2 amplitude distributions. We shall restrict our discussion here to the linear inversion scheme.

The NMR logging tools typically acquire T_2 CPMG spin-echo trains of several hundreds to a few thousands of echoes. The inversion process involved is to find a set of T_2 amplitudes, f_j, from a set of measurements of echoes, g_i:

$$g_i = M(t_i)/M_0 = \sum_{j=1}^{m} a_j f_j e^{-t_i/T_j} + \epsilon_i, \quad i = 1, ..., n \tag{6.10}$$

by minimizing the error ϵ_i. Here we assume that there are n echoes each with an amplitude g_i measured at t_i and m known (i.e., assumed) relaxation times T_j preselected to be equally spaced on a logarithmic scale. The echo amplitude g_i is given by the magnetization $M(t_i)$ of the system measured at t_i normalized by the magnetization at $t = 0$, M_0. In NMR logging data, the echo amplitudes are typically in porosity units (PU).

We have assumed that f_j is the partial porosity for the pores which have a T_2 relaxation time of value T_j associated with them. Note that we have a factor, a_j, besides f_j in Eq. (6.10). This factor a_j is the polarization factor, and is given by

$$a_j = 1 - e^{-RD/T_{1j}} \tag{6.11}$$

where RD is the repeat delay time or wait time. It is the time between successive exciting pulses for the CPMG echo sequences. T_{1j} is the T_1 relaxation time associated with these pores of size j.

Normally, we use a long enough wait time, typically three times the T_{1j} relaxation time to allow all the spins in the system to fully recover to their equilibrium state. In this situation, the factor a_j is one. Sometimes, we use a very short wait time purposely to measure only the short relaxation time components while suppressing the long relaxation time components. In this case, a_j will still be close to one for short time components, but can be suppressed to a fraction of one or near zero for long time components. When solving for f_j, this polarization factor is lumped into the factor e^{-t_i/T_j} by assuming a known T_{1j}/T_{2j} ratio (note that, in Eq. (6.10), T_{2j} is simplified as T_j). Thus, the polarization factor is given by

$$a_j = 1 - e^{-RD/rT_j} \tag{6.12}$$

where r is the T_{1j}/T_{2j} ratio. In the following discussion, we shall assume a_j is equal to 1.

Since all e^{-t_i/T_j}'s are known and we are solving for f_j, it is a linear inversion problem. A least squares fit is used to minimize the following sum:

$$\min\left\{\phi(f) = \sum_{i=1}^{n} \frac{1}{\sigma_i^2}(\sum_{j=1}^{m} f_j e^{-t_i/T_j} - g_i)^2\right\}, \tag{6.13}$$

where σ_i is the measurement error of the i-th echo, g_i.

This problem is well known to be *ill-posed.* Small fluctuations in the measured data can lead to drastically different sets of f_j's which all give reasonably good fits. It requires a procedure, known as *regularization*, to obtain a reasonable solution. To have a clear understanding of these terms, *ill-posed* and *regularization*, we shall discuss this further in the following section.

6.2.1 Ill posed problem

The subject of *ill-posed*, or *incorrectly-posed* problems has been studied extensively in the past by Tikhonov [211], Turchin [218], Phillips [175], Twomay [219], Wahba [221], and Butler et al. [48]. To define this in precise mathematical terms, we shall follow Tikhonov [211].

The problem $\mathcal{R}$ of the determination of a function $f(s) \in \mathcal{F}$ by means of a given function $g(t) \in \mathcal{G}$, $f(s) = \mathcal{R}[s, g(t)]$ (where $\mathcal{F}$ and $\mathcal{G}$ are function spaces), is said to be *correctly-posed* if:

1. For every function $g(t) \in \mathcal{G}$, there exists a solution $f(s)$ of the problem.

2. The solution $f(s)$ is unique for a given $g(t)$.

3. The dependence of the solution $f(s)$ on $g(t)$ is continuous in the metrics of $\mathcal{F}$ and $\mathcal{G}$

The problem $\mathcal{R}$ is said to be *incorrectly-posed* or *ill-posed*, if the above conditions cannot be satisfied. For example, in a *correctly-posed* problem, for a given function $\bar{g}(t)$, there corresponds a unique solution $\bar{f}(s) = \mathcal{R}[s, \bar{g}(t)]$ for the given problem $\mathcal{R}$. Moreover, even if we know $\bar{g}(t)$ only approximately, we can still determine $\bar{f}(s)$ reliably in the following sense: For any $\epsilon > 0$, no matter how small, there exists a $\delta > 0$ such that if $||g(t) - \bar{g}(t)|| < \delta$, then the exact unique solution $f(s) = \mathcal{R}[s, g(t)]$ satisfies $||f(s) - \bar{f}(s)|| < \epsilon$. In an *ill-posed* problem, by contrast, it is not possible, in general, to control the accuracy of $\bar{f}(s)$ in this fashion. In some cases, even the existence and/or uniqueness of $\bar{f}(s)$ is not ensured.

Many important practical problems are ill-posed. For instance, Fredholm's equation of the first kind is one of the most important examples. Eq. (6.10) is an example of a Fredholm equation of the first kind in discrete form. We can recast that in the following integral form with continuous variables and functions:

$$\hat{K}f \equiv \int_a^b K(t,s)f(s)ds = g(t), \qquad c \leq t \leq d \tag{6.14}$$

where e^{-t_i/T_j} is replaced by $K(t,s)$, f_j by $f(s)$, and g_i by $g(t)$, and s stands for the relaxation time T. The problem becomes to solve for $f(s)$, given $g(t)$ and the functional form of $K(t,s)$.

In this general form, $K(t,s)$ is called the kernel of the equation. It can be any arbitrary function, which determines the operator $\hat{K}$ of the problem and converts the unknown function $f(s)$ into some other function $g(t)$, which is accessible to physical measurements. The problem itself, in terms of the definitions above, is denoted as $\mathcal{R}$.

Frequently, the solution for Eq. (6.14) may not exist at all, or may not exist for every function $g(t)$ on the right hand side. For instance, if

$$K(t,s) = t + s \tag{6.15}$$

the right hand member $g(t)$ must be a linear function of t. If $g(t) = g_0 + g_1 t$, then any function $f(s)$ that satisfies the following conditions,

$$\int_a^b f(s)ds = g_1, \qquad \int_a^b sf(s)ds = g_0, \tag{6.16}$$

is a solution of Eq. (6.14) with kernel (6.15). It is clear that the solution for $f(s)$ is not unique, and the problem is consequently *ill-posed.*

Sometimes, in problems arising from physics, we are fairly sure of the existence and uniqueness of the function $f(s)$ (e.g., the inverse problem of potential theory). However, even in this case, Eq. (6.14) possesses a peculiar property which makes the problem an *ill-posed* one. This peculiar property is the "smoothing" action of the kernel. We shall illustrate this with an example.

Let the kernel $K(t,s)$ be a continuous function of its second argument s. Let us assume that there exists a solution $f_1(s)$ for a given $g_1(t)$. Then the function $f_2(s) = f_1(s) + C \sin 2\pi ns/(b-a)$, where C is a large number and n is a sufficiently large integer, can also be made an approximate solution of the problem. It is quite clear that for any arbitrarily large C, we can find a large enough n such that the absolute value of the difference of the corresponding right hand members $g_1 = \hat{K} f_1$ and $g_2 = \hat{K} f_2$ can be made arbitrarily small. Thus the kernel in question has "*smoothed out*" a very intense, but very high frequency component to an extremely small level. If the right hand member $g(t)$ could be measured exactly, then the uniqueness of the solution $f(s)$ would have been guaranteed. However, measurement error always exists in obtaining $g(t)$. Thus wildly different $f(s)$, such as $f_1(s)$ and $f_2(s)$ always exist, and in fact, there are infinitely many such solutions. The problem has become *ill-posed.*

In practical physical problems, such as the NMR data inversion discussed here, the numerical solution of Eq. (6.14) always reduces it in one way or another to a system of linear algebraic equations

$$(\hat{K} f)_i \equiv \sum_{j=1}^{m} K_{ij} f_j = g_i, \quad i = 1, 2, \ldots, n \tag{6.17}$$

where K_{ij}, f_j, and g_i are discrete values of their respective continuous functions, and g_i's are accessible for measurements while f_j are to be determined.

Thus the uniqueness of the solution depends strongly on the precise measurement of g. Any error in g can lead to drastically different $f(s)$. It has been shown (Fadeev [81], Lanczos [131]) that the stability of the solution for such an inversion problem is closely related to the set of eigenvalues μ of the matrix $K^T K$, where K and K^T are the matrix forms of the operator $\hat{K}$ and the adjoint operator $\hat{K}^\dagger$, respectively. The stability decreases as the ratio $\mu_{\max}/\mu_{\min}$ increases. Consequently, the demand for measurement precision and accuracy for $g(t)$ becomes greater and greater. The measurement error propagates to all components of the solution $f(s)$ through a factor proportional to μ^{-1}. Thus as the μ we include in the inversion gets smaller, the solution becomes more unstable. We shall discuss this further in the section on singular value decomposition.

Faced with such a situation, it is really hopeless mathematically, unless we can inject new information into the process and place some kind of physical constraints on the solution $f(s)$ based on a priori knowledge about the nature of $f(s)$. One such constraint is the belief that the function $f(s)$ should be reasonably smooth. To incorporate this into the inversion process, the general approach is to add a penalty function in Eq. (6.13) to smooth or *regularize* the solution. This *regularization* process can be defined more precisely as follows.

We shall call a process $\mathcal{R}_\delta$ a regularizing process (or algorithm) if :

1. $\mathcal{R}_\delta[s, \bar{g}(t)]$ is defined for every $\bar{g} \in \mathcal{G}$ and $\delta > 0$.

2. If for $\bar{g}(t)$ there exists a $\bar{f}(s) = \mathcal{R}[s, \bar{g}(t)]$, then for every positive ϵ there exists a $\delta(\epsilon, \bar{f})$ such that if $\|\tilde{g}(t) - \bar{g}(t)\|_{\mathcal{G}} < \delta$, then $\|\tilde{f}_\delta(s) - \bar{f}(s)\|_{\mathcal{F}} \leq \epsilon$, where $\tilde{f}_\delta = \mathcal{R}_\delta[s, \tilde{g}]$.

We shall say that a problem $f(s) = \mathcal{R}[s, g(t)]$ is a *regularizable* problem if it has at least one regularizing process.

Here we introduce three types of *regularization* depending on the types of penalty function, namely: the "norm" (or the "amplitude") smoothing,

$$\min\left\{\phi(f) = \frac{1}{2}\sum_{i=1}^{n}(\sum_{j=1}^{m} f_j k_{ij} - g_i)^2 + \frac{\alpha}{2}\sum_{j=1}^{m} f_j^2\right\} \tag{6.18}$$

the "slope" smoothing,

$$\min\left\{\phi(f) = \frac{1}{2}\sum_{i=1}^{n}(\sum_{j=1}^{m} f_j k_{ij} - g_i)^2 + \frac{\alpha}{2}\sum_{j=1}^{m}(f_{j+1} - f_j)^2\right\} \tag{6.19}$$

and the "curvature" smoothing,

$$\min\left\{\phi(f) = \frac{1}{2}\sum_{i=1}^{n}(\sum_{j=1}^{m} f_j k_{ij} - g_i)^2 + \frac{\alpha}{2}\sum_{j=1}^{m}(f_{j+1} + f_{j-1} - 2f_j)^2\right\} \tag{6.20}$$

where $k_{ij} = e^{-t_i/T_j}$, and α, often referred to as a "smoothing" or a "regularization" parameter, is a constant optimally chosen to be commensurate with the measurement error. Recent work by Borgia, et al [34] extends the treatment on the "curvature" smoothing to provide a uniform penalty along the relaxation time axis by allowing the constant α to be a variable.

Other regularizing processes such as the basis function method and singular value decomposition method do not specifically have a penalty function. However, there are also regularizing actions involved in the selection of basis functions or singular values. Although there are many regularization processes for the NMR CPMG data inversion problem, the characteristic shape of the T_2 distributions obtained by various methods are by and large similar, and the total sums of the amplitudes are generally in agreement. For a specific regularization process, the measurement error δ determines the uncertainty of the solution ϵ. There may be slight difference in the shape of the T_2 distributions (i.e. $f(s)$). This is because different regularization processes impose different constraints on the infinitely many possible solutions $f(s)$.

6.2.2 Norm smoothing

There are multiple methods for implementing "norm" smoothing. One commonly used procedure for minimizing Eq. (6.18) is the Levenberg-Marquardt method [180] with the non-negative constraint for all f_j's in the iterative process. The general concern, however, is how to choose an optimal value for the smoothing parameter α to be commensurate with the measurement error.

A frequently used criterion is to minimize the difference between the vector $f^{(\alpha)}$ (the α-dependent solution from Eq. (6.18)) and $f^{(0)}$ (the true solution), and somehow relate that to the known measurement error of the echo amplitudes. With the input of the known measurement error, the optimal value for α is then determined.

Here, we shall present a method proposed by Butler, Reeds, and Dawson (BRD) [48]. The following discussion is based on a simplified description by Bergman [15] of the original Butler-Reeds-Dawson paper.

The problem is to minimize the following quantity

$$\phi(f_x) = \phi(f_1 \dots f_m) \equiv \frac{1}{2}\sum_{i=1}^{n}\left(\sum_{x=1}^{m} k_{ix}f_x - g_i\right)^2 + \frac{\alpha}{2}\sum_{x=1}^{m} f_x^2. \qquad (6.21)$$

by finding a set of f_x which satisfy the positivity constraint

$$f_x \geq 0 \text{ for all } x.$$

Eq. (6.21) is minimized when the following set of $n+m$ coupled equations are satisfied by the $n+m$ unknowns $f_x^{(\alpha)}$, c_i

$$\sum_{y=1}^{m} k_{iy} f_y^{(\alpha)} - g_i + \alpha c_i = 0 \quad , \qquad i = 1 \dots n \qquad (6.22)$$

$$f_x^{(\alpha)} = \max\left(0, \sum_{i=1}^{n} c_i k_{ix}\right) \quad , \qquad x = 1 \dots m. \qquad (6.23)$$

To show this we note that the constrained minimum of $\phi(f_x)$ can be found by considering its m-dimensional gradient, given by

$$(\nabla\phi)_x \equiv \frac{\partial\phi}{\partial f_x} = \sum_{i=1}^{n} k_{ix}\left(\sum_{y=1}^{m} k_{iy}f_y - g_i\right) + \alpha f_x. \tag{6.24}$$

At the calculus minimum, all components of $\nabla\phi$ must vanish. However, if some of the f_x happen to be negative, then the *constrained* minimum will have some of these components (not necessarily the same components that are negative at the calculus minimum!) set to their extreme allowed nonnegative values, namely 0. At such a *non-calculus* minimum, the components of $\nabla\phi$ whose corresponding f_x were made 0 will be *positive*, while the other components of $\nabla\phi$ whose corresponding f_x were positive, vanish. Hence at the constrained minimum

$$\begin{aligned}(\nabla\phi)_x &= 0 \quad \text{for} \quad f_x > 0\\ (\nabla\phi)_x &\geq 0 \quad \text{for} \quad f_x = 0.\end{aligned} \tag{6.25}$$

The gradient of ϕ can be calculated for $f_x = f_x^{(\alpha)}$, which are solutions of (6.22) and (6.23): We multiply (6.22) by k_{ix}, sum over i, and obtain

$$0 = \sum_{i=1}^{n} k_{ix}\left(\sum_{y=1}^{m} k_{iy}f_y^{(\alpha)} - g_i\right) + \alpha\sum_{i=1}^{n} c_i k_{ix}. \tag{6.26}$$

If $f_x > 0$, then the last term in this equation is equal to αf_x, and we get $(\nabla\phi)_x = 0$. If $f_x = 0$, then the last term in this equation must be negative, and (6.23)–(6.26) imply that $(\nabla\phi)_x \geq 0$. This proves that the solution of (6.22) and (6.23) minimizes (6.21) and satisfies the positivity constraint (6.25). To solve the coupled equations (6.22) and (6.23), we note that they can be modified to the following set of n equations for the c_i

$$\sum_{j=1}^{n} M_{ij}c_j + \alpha c_i = g_i, \tag{6.27}$$

where the matrix $\hat{M}$ is defined by

$$M_{ij} \equiv {\sum_x}' k_{ix}k_{jx}. \tag{6.28}$$

and $\sum_x'$ stands for a restricted sum over those values of the index x for which $\sum_{i=1}^{n} c_i k_{ix} > 0$. Because of this restriction, the matrix $\hat{M}$ depends upon the c_i, and (6.27) is *not* a set of linear equations.

Eqs. (6.27) can be solved in a number of ways. One way is to note that the solution also minimizes the following quantity

$$\psi \equiv \frac{1}{2} c^\dagger (\hat{M} + \alpha \hat{I}) c - c \cdot g, \tag{6.29}$$

and proceed to find a vector c at which ψ is a minimum. Another method for solving (6.27) is to start with a full matrix $M_{ij} = \sum_x k_{ix} k_{jx}$ without the restriction on the summation, solve for c, then update M_{ij} according to (6.28), and iterate to the final correct solution.

The problem of optimizing the regularization parameter α is handled by attempting to minimize the difference between the vector $f^{(\alpha)}$ and $f^{(0)}$, i.e.

$$(f^{(\alpha)} - f^{(0)})^\dagger \cdot (f^{(\alpha)} - f^{(0)}) = \sum_x (f_x^{(\alpha)} - f_x^{(0)})^2 \tag{6.30}$$

where $f^{(0)}$ is the "true" f being sought. Since $f^{(0)}$ is independent of α, this is equivalent to minimizing

$$\sum_x (f_x^{(\alpha)} - f_x^{(0)})^2 - \sum_x (f_x^{(0)})^2 = \sum_x (f_x^{(\alpha)})^2 - 2 \sum_x f_x^{(\alpha)} f_x^{(0)}. \tag{6.31}$$

We then develop the last, unknown term as follows

$$\sum_x f_x^{(\alpha)} f_x^{(0)} = \sum_i {\sum_x}' c_i k_{ix} f_x^{(0)} \geq \sum_i c_i g_i^{(0)}, \tag{6.32}$$

where $g_i^{(0)}$ denotes the ideal spin-echo amplitudes with no noise. If we denote the noise by ϵ_i, we finally get the following upper bound for (6.31), using (6.23) and (6.28),

$$c^\dagger \hat{M} c - 2c \cdot g^{(0)} = c^\dagger \hat{M} c - 2c \cdot g + 2c \cdot \epsilon, \tag{6.33}$$

where $g_i = g_i^{(0)} + \epsilon_i$.

This quantity, which depends on α (through $\hat{M}$ and c_i) and on ϵ_i, must be minimized by varying α over positive values. The lack of knowledge about $f_x^{(0)}$ has been translated into a lack of knowledge about ϵ_i, which is more amenable to estimations.

Assuming that the ϵ_i have the character of random noise, i.e., that they are independently distributed random variables with zero mean and identical variances σ^2, a reasonably conservative estimate of the "maximum likely value" of $c \cdot \epsilon$ is

$$\sqrt{n}(c \cdot c)^{1/2} \sigma.$$

When this is substituted in (6.33), the resulting expression

$$H(\alpha) \equiv c^{\dagger}\hat{M}c - 2c \cdot g + 2\sqrt{n}\sigma(c \cdot c)^{1/2} \tag{6.34}$$

is entirely determined by the solution of (6.27). The function $H(\alpha)$ has a derivative given by

$$\frac{1}{2}H'(\alpha) = c^{\dagger}(\hat{M} + \alpha\hat{I})^{-1}c\left(\alpha - \frac{\sqrt{n}\sigma}{(c \cdot c)^{1/2}}\right),$$

the sign of which is determined by whether

$$\xi(\alpha) \equiv \alpha^2(c \cdot c) = \alpha^2 g^{\dagger}(\hat{M} + \alpha\hat{I})^{-2}g = (g - \hat{M}c)^{\dagger} \cdot (g - \hat{M}c) > 0 \tag{6.35}$$

is greater or smaller than the constant value $n\sigma^2$. Note that $\alpha c = g - \hat{M}c$, where $\hat{M}c = \sum_x k_{ix} f_x$ represents all vectors expressible as such a summation with some values of f_x.

Using the derivative of $\xi(\alpha)$, given by

$$\xi'(\alpha) = \alpha c \cdot c - \alpha^2 c^{\dagger}(\hat{M} + \alpha\hat{I})^{-1}c = \alpha c^{\dagger}(\hat{M} + \alpha\hat{I})^{-1}\hat{M}c > 0, \tag{6.36}$$

it is easy to construct a simple procedure for finding where $\xi(\alpha) = n\sigma^2$, which is a unique calculus minimum of $H(\alpha)$ if it exists, and provides the optimum value of α, α_{opt}.

In general, since $\xi(\alpha)$ is a monotonically increasing function, the existence of such a finite, positive value for α is assured if

$$\xi(0) < n\sigma^2 < \xi(\infty). \tag{6.37}$$

From Eq. (6.35) we get

$$\xi(\infty) = g \cdot g, \quad \xi(0) = (g - g^*) \cdot (g - g^*),$$

where g^* is the closest point to g in the space of all n-dimensional vectors expressible as $\sum_x k_{ix} f_x$ for some values of f_x. In the usual case, when

$$(g - g^*) \cdot (g - g^*) < n\sigma^2 < g \cdot g,$$

Eq. (6.37) is satisfied and a unique, finite value for α_{opt} exists and can be found using the above mentioned procedure (i.e., a procedure which finds an α_{opt} such that $\xi(\alpha_{opt}) = n\sigma^2$). If the signal-to-noise ratio is too small, so that $n\sigma^2 > g \cdot g$, we get $\alpha_{opt} = \infty$ and a meaningless, flat distribution. If $(g - g^*) \cdot (g - g^*) > n\sigma^2$, our choice of relaxation time spectrum was bad since it is incapable of reproducing the measured spin-echo amplitudes to a good approximation, and we get $\alpha_{opt} = 0$. If this occurs, it means that we should either improve our measurement accuracy or use a different model.

There are at least two ways of implementing the BRD method. The first one suggested by Bergman [15] uses the fact that the elements of $\hat{M}$ are piecewise constant when the c_i's change, and it works as follows:

(1) For a fixed α, we find a solution c which satisfies

$$(M_{ij} + \alpha\delta_{ij})c_j = g_i \tag{6.38}$$

where $M_{ij} = \sum_{x=1}^{m} k_{ix}k_{jx}$.

(2) The f_x is then given by

$$f_x = \max(0, \sum_{i=1}^{n} c_i k_{ix}). \tag{6.39}$$

(3) We then update the matrix element $M_{ij} = \sum_x' k_{ix}k_{jx}$. This time it is summed over only those x's which give positive values for $\sum_{i=1}^{n} c_i k_{ix}$. The new M_{ij} is used in (1) to solve for a new c.

(4) Repeat the process, until c stops changing. The final value for f_x is given by (2).

This procedure usually works quite well. However, it is possible that direct iterations of Eq. (6.38) to solve for c may be impossible. In that case, the second procedure which was suggested by Butler et al. should be used, i.e., to search for the minimum of the convex function

$$\psi = \frac{1}{2}c^\dagger(\hat{M} + \alpha\hat{I})c - c \cdot g \tag{6.40}$$

This can be accomplished by evaluating

$$\psi' \equiv \frac{\partial\psi}{\partial c_i} = \sum_j (\hat{M} + \alpha\hat{I})_{ij}c_j - g_i \tag{6.41}$$

$$\psi'' \equiv \frac{\partial^2\psi}{\partial c_i \partial c_j} = (\hat{M} + \alpha\hat{I})_{ij} \tag{6.42}$$

and obtain a new c as

$$c(\text{new}) = c(\text{old}) - \gamma\Delta, \tag{6.43}$$

where $\Delta = \psi'/\psi''$ and γ is the first of $(\frac{1}{2})^0, (\frac{1}{2})^1, (\frac{1}{2})^2, \cdots$, such that

$$\psi(c(\text{new})) < \psi(c(\text{old})). \tag{6.44}$$

The search is typically terminated when

$$\frac{\| (\hat{M} + \alpha\hat{I})c - g \|}{\| g \|} \leq 10^{-6}, \tag{6.45}$$

an arbitrary selected tolerance.

To search for the optimal α, one uses the fact that $\xi(\alpha)$ is a monotonic increasing function and designs a simple scheme to obtain

$$\xi(\alpha_{opt}) = \alpha_{opt}^2(c \cdot c) = n\sigma^2. \tag{6.46}$$

There are many other "norm" smoothing methods. One of such methods is described by Freedman [88].

6.2.3 Curvature smoothing

The smoothing parameter α for *curvature smoothing* is determined empirically by plotting the standard deviation of the fit as a function of α. The optimal α is chosen where the rate of increase of the standard deviation starts to increase dramatically. However, there seems to be no satisfactory way of handling noisy data using the curvature smoothing method. When applied to noisy data, much larger α is often required to get the right porosity, which results in an oversmoothed T_2 distribution.

One of the common procedures in the application of *curvature smoothing* is to use the Levenberg-Marquardt method to minimize Eq. (6.20) [180]. The idea is to "*walk*" from the point of initial guess for f to the point where the function ϕ of Eq. (6.20) is a minimum. This can be accomplished by assuming ϕ is quite close to the minimum and has a quadratic form as follows:

$$\phi(f + \Delta f) = \phi(f) - \beta \cdot \Delta f + \frac{1}{2}\Delta f^{\dagger} \cdot \omega \cdot \Delta f \tag{6.47}$$

where

$$\beta_x = -\frac{\partial \phi}{\partial f_x}, \qquad \omega_{xy} = \frac{\partial^2 \phi}{\partial f_x \partial f_y}, \tag{6.48}$$

and $f + \Delta f$ represents the final correct solution, f the solution before it reaches the final answer. At the point of correct solution, we have

$$\frac{\partial \phi(f + \Delta f)}{\partial \Delta f} = 0 \tag{6.49}$$

which leads to

$$\omega \cdot \Delta f = \beta. \tag{6.50}$$

From (6.50), we can solve for Δf . This is called the inverse-Hessian method. This would have taken us to the minimum point of ϕ if we were very close to the minimum. But the function is often not exactly quadratic and the initial guess for f can be far from the minimum. The safest way to numerically "*descend*" to a lower value for ϕ is to "*walk*" along the negative gradient of ϕ, i.e.,

$$\Delta f_x = \text{constant} \times \beta_x, \tag{6.51}$$

which is a method known as *steepest descent.* The Levenberg-Marquardt method provides a smooth switching between the extremes of the inverse-Hessian method and the steepest descent method by introducing a λ parameter as follows:

$$\omega'_{xy} = \omega_{xy}(1 + \lambda \delta_{xy}). \tag{6.52}$$

The Δf_x is solved from

$$\omega'_{xy} \Delta f_y = \beta_x, \qquad f_x(\text{new}) = f_x(\text{old}) + \Delta f_x. \tag{6.53}$$

We then use the new f_x to evaluate ϕ of (6.20). Initially, the λ is set to an arbitrarily small number. If the new value for ϕ is greater than the previous value for ϕ, we increase λ by a factor of 10 and repeat the process until the new ϕ becomes less than the old ϕ. (By increasing λ, we are limiting the step size of the "*walk*", converging along the steepest descent, and ensuring a decreasing ϕ.) When this is satisfied, the new f_x is accepted, and the process is repeated to find the new f_x by solving (6.53) with a new λ decreased by a factor of 10. This pushes the solution towards the inverse-Hessian method and speeds up the iteration process. To ensure the non-negativity requirement for f_x during the iterations, any new f_x which is negative is set to zero, before the evaluation of ϕ. The iteration process stops when the fractional change of ϕ is less than a selected tolerance, typically 10^{-5}.

6.2.4 Uniform penalty method

Recently, Borgia, Brown, and Fantazzini [34] devised an elaborate scheme for "curvature smoothing" by allowing the regularization parameter "α" to be a variable to provide a uniform penalty (UPEN) along the relaxation time axis. A feedback mechanism is built in such that the weight of the penalty term approaches a uniform value throughout all the relaxation times. The net effect of such a feedback is to make sharp peaks sharper and smooth bumps smoother along the T_2 relaxation time axis.

UPEN uses some global smoothing, or regularizing parameters, which are normally not changed from sample to sample. It uses several computed noise parameters in the regularization process, however. It also allows data windowing, with corresponding weighting factors, for T_2 data sets with large numbers of echoes.

The ability of UPEN to fit relaxation data corresponding to distributions containing both sharp and broad features makes it possible in many cases to identify systematic errors in data that can lead to an inappropriate interpretation. For instance, spurious resolved peaks are sometimes produced by data distortion corresponding to a slight downward bending of the logarithmic decay curve. Non-random errors of fit are likely to result, and in many cases there is a significant *cost* for application of the usual non-negativity constraint, in which case the program posts a warning.

With good relaxation data the non-negativity constraint is usually not needed in UPEN if there are no peaks narrow enough to have only a few computed points and if there are no very sharp valleys. By *good relaxation data* is meant data representing a sum of positive exponential components, with random noise added. Most good petrophysical relaxation data do not require the non-negativity constraint with UPEN.

An awkward feature of UPEN is due to the need to include in the feedback process the nearest neighbors of each point on the computed distribution. This can cause the iterative procedure not to converge. In such cases, it usually wanders around a small region of solution space where changes are not significant for interpretation purposes. A smooth procedure for promoting convergence is implemented, if needed after many iterations, by retaining the smaller of the last two penalty values. This brings rapid convergence without significantly changing the computed distribution. In most cases, the iterations converge without this procedure.

6.2.5 Basis function method

In the simple "norm", "slope", and "curvature" smoothing methods, the penalty terms are added to suppress the oscillatory behavior of the unknown function $f(s)$. However, such suppression is not always successful especially for high signal-to-noise data where the high precision of the measurements of $g(t)$ allows smaller eigenvalues to enter the inversion process. We can either use a regularizing process, such as uniform penalty method, with a specific feedback algorithm to suppress the oscillatory behavior of $f(s)$, or use a basis function with a broad base to exclude solutions with rapid oscillatory features.

In solving the Fredholm equation of the first kind in discrete form, such as Eq. (6.17), we have assumed $K_{ij} = e^{-t_i/T_j}$ and solved for the corresponding amplitude f_j, in other words, a single amplitude f_j for a single relaxation time T_j. The determination of f_j does not directly affect the amplitude of neighboring T_j's, i.e. T_{j-1} and T_{j+1}, etc. This is the reason why for high signal-to-noise data, it becomes difficult to suppress the oscillatory behavior of f. To prevent this from happening, one can couple the determination of f_j to the those of the neighboring T_js. One simple example is that f_j, instead of being the amplitude for e^{-t/T_j}, becomes the amplitude for

$$\frac{1}{2}e^{-t/T_{j-1}} + e^{-t/T_j} + \frac{1}{2}e^{-t/T_{j+1}} \tag{6.54}$$

Thus the determination of f_j has an effect on the amplitudes of the neighboring relaxation times, and vice versa. Having a basis function to represent each f_j forces the solution to have a smooth behavior.

Eq. (6.54) is just an example for the basis function. Watson et al. [142] use the B-splines as basis functions, where the magnetization $M(t_i)/M_0$ is expressed as:

$$M(t_i)/M_0 = g_i = \sum_{j=1}^{m} C_j \int_0^{\infty} B_j^p(T) e^{-t_i/T} dT \tag{6.55}$$

The regularization process is carried out by minimizing the following expression:

$$\min\left\{\sum_{i=1}^{n}(\sum_{j=1}^{m} C_j \int B_j^p(T)e^{-t_i/T}dT - g_i)^2 + \lambda \sum_{j=1}^{m}(C_{j+1} + C_{j-1} - 2C_j)^2\right\} \tag{6.56}$$

where curvature smoothing is used.

In fact, if the basis function is broad enough, the *regularizing* action is already included. A penalty term is not needed. The smoothing of the solution f, is to some degree, determined by the broadness of the basis function. Recently, Miller et al. [162] has shown that using gamma functions as basis functions can lead to a simple analytic form for the linear algebraic equations.

The magnetization $M(t_i)/M_0$ is expressed as :

$$M(t_i)/M_0 = g_i = \sum_{j=1}^{m} A_j \int F_j(T)e^{-t_i/T}dT = \sum A_j I_j \tag{6.57}$$

where the basis function is given by:

$$F_j(s) = (\kappa s/s_j)^{\kappa} \exp(-\kappa s/s_j)/(s\Gamma(\kappa)) \tag{6.58}$$

where $s = T^{-1}$, and κ is a constant which determines the broadness of the basis function. As κ increases, the spread of the basis function decreases.

This choice of the basis functions results in a simple analytic form for the integrals, I_j;

$$I_j(t_i) = \left(1 + \frac{s_j t_i}{\kappa}\right)^{-\kappa} \tag{6.59}$$

which fits in nicely in a set of linear algebraic equations.

6.2.6 Singular value decomposition

One of the popular methods for treating the linear inversion problem is the "*singular value decomposition*" (SVD) method. The SVD method is particularly attractive for the following reasons: (1) it does not require a penalty term for smoothing or regularization, thus obviating the need to search for an optimum regularization parameter, (2) it provides a clear picture of the characteristics of the matrix, as well as the overall properties of the inversion procedure, and (3) it offers a straightforward and easily comprehensible method for obtaining a "regularized" solution by removing the small singular values in a fashion that reflects the measurement error.

In general, we assume that the measurement error is the same for every spin echo. Thus, σ_i in Eq. (6.13) is the same for all i. However, the procedure becomes somewhat unclear when we need to process simultaneously two or more NMR data sets, with very different measurement errors, coming from the same sample. In fact, one encounters a similar problem when the

NMR time domain data are compressed or averaged within different window lengths. This windowing leads to different improved precisions.

In the following, we take a closer look at the procedure used in the SVD inversion as to how the noise level affects the solution. A geometrical interpretation, which provides an intuitive picture for some of the formal mathematical considerations described below, can be found in Ref. [79].

Without loss of generality, we shall limit the following discussion to the simultaneous processing of two data sets with different measurement errors. Suppose we need to minimize the following two sets of data simultaneously:

$$\min\{\sum_{i=1}^{p}\frac{1}{\sigma_1^2}(\sum_{j=1}^{m}k_{ij}y_j-b_i)^2\} \tag{6.60}$$

$$\min\{\sum_{i=1}^{q}\frac{1}{\sigma_2^2}(\sum_{j=1}^{m}l_{ij}y_j-c_i)^2\} \tag{6.61}$$

where

$$k_{ij} = (1-e^{-RD_1/rT_j})e^{-t_i/T_j} \tag{6.62}$$

$$l_{ij} = (1-e^{-RD_2/rT_j})e^{-t_i/T_j}; \tag{6.63}$$

RD_1 and RD_2 represent the RD (repeat delay time or recovery time) of runs 1 and 2; σ_1 and σ_2, the corresponding measurement errors; b_i and c_i, the corresponding echo trains; and p and q, the corresponding number of echoes (or data points) in the respective runs. We want to solve both data sets simultaneously so that they yield the same solution y_j. We shall call this procedure the "*composite-data processing.*"

To set up the combined data sets as one least squares fit problem, we write

$$\begin{bmatrix} A_{11} & A_{12} & \dots & A_{1m} \\ \vdots & & & \vdots \\ \vdots & & & \vdots \\ A_{n1} & A_{n2} & \dots & A_{nm} \end{bmatrix}\begin{bmatrix} y_1 \\ \vdots \\ \vdots \\ y_m \end{bmatrix}$$

$$= \begin{bmatrix} k_{11} & k_{12} & \dots & k_{1m} \\ \vdots & & & \vdots \\ \frac{\sigma_1}{\sigma_2}l_{11} & \frac{\sigma_1}{\sigma_2}l_{12} & \dots & \frac{\sigma_1}{\sigma_2}l_{1m} \\ \vdots & & & \vdots \end{bmatrix}\begin{bmatrix} y_1 \\ \vdots \\ \vdots \\ y_m \end{bmatrix} = \begin{bmatrix} b_1 \\ \vdots \\ \frac{\sigma_1}{\sigma_2}c_1 \\ \vdots \end{bmatrix} = \begin{bmatrix} b_1 \\ \vdots \\ \vdots \\ b_n \end{bmatrix} \equiv \mathsf{b}, \tag{6.64}$$

where A_{ij} are the elements of the combined $m \times n$ matrix A with $n = p + q$, and b is the combined data vector with n elements.

Note that the data for run 2, as well as the corresponding matrix elements, have now been scaled by a factor of σ_1/σ_2, so that the combined data set has a common noise level of σ_1. The latter will be used to determine the cutoff in the SVD inversion procedure.

Once the combined data sets are set up as one least squares fit problem with the proper weighting taken care of, the method for solving the equation need not to be limited to the SVD technique. The combined data set need not be restricted to two data sets, either.

In fact, we encounter a similar situation of different variances when compressing the time domain data into several windows to increase the computation speed. For example, Eq. (6.13) can be transformed to

$$\sum_{i=1}^{s} \frac{1}{N_i \sigma^2} (\sum_{j=1}^{m} f_j K_{ij} - B_i)^2, \tag{6.65}$$

where n echoes have been partitioned into s windows ($i = 1, \ldots, s$) with N_i echoes in the i-th window, and

$$\begin{aligned} n &= N_1 + \ldots + N_{i-1} + N_i + \ldots + N_s, \\ r_i &= N_1 + \ldots + N_{i-1}, \quad r_1 = 0, \\ K_{ij} &= \sum_{k=r_i+1}^{r_i+N_i} e^{-t_k/T_j}, \\ B_i &= \sum_{k=r_i+1}^{r_i+N_i} b_k. \end{aligned} \tag{6.66}$$

Now the variance for B_i is $N_i \sigma^2$, and each window has a different variance.

Similarly, we can also average the data within each window, assuming that arithmetic averaging is a valid procedure for this type of data, to obtain

$$\sum_{i=1}^{s} \frac{1}{(\sigma/\sqrt{N_i})^2} (\sum_{j=1}^{m} f_j e^{-t_i/T_j} - \bar{b}_i)^2, \tag{6.67}$$

where $\bar{b}_i$ is simply the average of the data within the i-th window, and t_i is an appropriate mid-point within that window. The noise level for the i-th window is now reduced to $\sigma/\sqrt{N_i}$.

Regardless of time domain data sum or average, it is easy to rescale the matrix elements and the data vectors so that all windows have the same noise level, as we did in the composite-data processing. This common noise level can then be used to determine the cutoff in the SVD inversion procedure.

Now that a common noise level for a composite-data set is established, we need to determine how to cut off the singular values in the SVD inversion process. The singular value decomposition theorem in linear algebra [131, 140, 180] states that any real $n \times m$ matrix A $(n \geq m)$ can be written as the product of an $n \times n$ orthonormal matrix U, an $n \times m$ diagonal matrix Λ with positive or zero diagonal elements, and the transpose of an $m \times m$ orthonormal matrix V, i.e.

$$\mathsf{A} = \mathsf{U}\Lambda\mathsf{V}^{\mathsf{T}} \tag{6.68}$$

where

$$\mathsf{U}^{\mathsf{T}}\mathsf{U} = \mathsf{I}, \quad \mathsf{V}^{\mathsf{T}}\mathsf{V} = \mathsf{I} \tag{6.69}$$

$$\sum_{i=1}^{n} U_{ip}U_{iq} = \delta_{pq}, \quad 1 \leq p, q \leq n \tag{6.70}$$

$$\sum_{i=1}^{m} V_{ip}V_{iq} = \delta_{pq}, \quad 1 \leq p, q \leq m \tag{6.71}$$

and

$$\Lambda = \mathrm{diag}(\lambda_1, \lambda_2, ..., \lambda_m). \tag{6.72}$$

The diagonal elements $\lambda_1 > \lambda_2 > ... > \lambda_m \geq 0$ are called the "singular values" of the matrix A.

The straightforward Householder transformations used for the decomposition of A need not produce all positive singular values λ's in descending order. However, it is easy to make them all positive by changing the signs of those that are negative and changing the signs of the corresponding eigenvectors, i.e., either the column vectors u_i of U or the column vectors v_i of V. It is also straightforward to rearrange the singular values in a descending order by rearranging the corresponding columns of U and V. Once the program is written in such a way as to produce a Λ with all the singular values positive and in descending order, the information about the ordering of the eigenvectors is already contained in both U and V matrices.

Eq. (6.64) can be written concisely, using vector/matrix notation, as

$$\mathsf{Ay} = \mathsf{b} \tag{6.73}$$

where the matrix $\mathsf{A} \in \Re^{n\times m}$ operates on a vector $\mathsf{y} \in \Re^m$ to produce a vector $\mathsf{b} \in \Re^n$, and A has all the properties described in Eqs. (6.68-6.72). Then

$$||\mathsf{Ay}||^2 = \mathsf{y}^\mathsf{T}\mathsf{A}^\mathsf{T}\mathsf{Ay} = \mathsf{y}'^\mathsf{T}\Lambda^\mathsf{T}\Lambda\mathsf{y}' = ||\Lambda\mathsf{y}'||^2 \tag{6.74}$$

where $\mathsf{A} = \mathsf{U}\Lambda\mathsf{V}^\mathsf{T}$, $\mathsf{y} = \mathsf{V}\mathsf{y}'$, and $||\ ||^2$ represents the squared norm of a vector, i.e., the sum of squares of all its components.

The largest (smallest) value of $||\mathsf{Ay}||^2/||\mathsf{y}||^2$ occurs when y lies along the direction of the eigenvector of the square orthogonal matrix V (actually, a column of that matrix) corresponding to the largest (smallest) singular value λ_1 (λ_m). Hence, for an arbitrary vector y, the following inequality is established:

$$\lambda_1^2 \geq \frac{||\mathsf{Ay}||^2}{||\mathsf{y}||^2} \geq \lambda_m^2, \tag{6.75}$$

which means, for the vector y which solves Eq. (6.73):

$$\lambda_1^2 \geq \frac{||\mathsf{b}||^2}{||\mathsf{y}||^2} \geq \lambda_m^2. \tag{6.76}$$

Frequently, we measure b and try to determine y. There is always measurement error involved in obtaining b. Therefore, instead of b and y, we have $\mathsf{B} = \mathsf{b} + \delta\mathsf{b}$ and $\mathsf{Y} = \mathsf{y} + \delta\mathsf{y}$ and are faced with the problem of solving

$$\mathsf{A}(\mathsf{y} + \delta\mathsf{y}) = \mathsf{b} + \delta\mathsf{b} \tag{6.77}$$

rather than Eq. (6.73): $\mathsf{Ay} = \mathsf{b}$. We note that y and b represent, respectively, the ideal solution and the ideal data vector without measurement errors, while $\delta\mathsf{y}$ and $\delta\mathsf{b}$ represent, respectively, the deviation in the solution and the measurement error of the data vector.

In order to satisfy both Eqs. (6.77) and (6.73), we must have

$$\mathsf{A}\delta\mathsf{y} = \delta\mathsf{b}. \tag{6.78}$$

Assuming that the matrix A is of rank m, its form in the principal coordinates or principal axes is represented by Λ, with m non-vanishing descending singular values $\lambda_1 \ldots \lambda_m$, and the normalized left and right singular vectors are denoted by $\mathsf{u}_1 \ldots \mathsf{u}_\mathsf{n}$, and $\mathsf{v}_1 \ldots \mathsf{v}_\mathsf{m}$, respectively. We can then transform y and b to the principal axes as y' and b':

$$\mathsf{y}' = \mathsf{V}^\mathsf{T}\mathsf{y} \tag{6.79}$$

$$\mathsf{b}' = \mathsf{U}^\mathsf{T}\mathsf{b}, \tag{6.80}$$

which replaces Eq. (6.91), is ensured. Once a proper cutoff such as Eq. (6.93) is determined, both the variances and the covariances of the uncertainties δy_i can be calculated using Eq. (6.86).

The singular values smaller than λ_r, (i.e., all λ_i, $i > r$) which are discarded in the restricted sums, are effectively replaced by 0 in the inverse matrix $\mathsf{\Lambda}^{-1}$. This means that the matrix $\mathsf{\Lambda}^{-1}$ is replaced by a matrix of lower rank r, $\mathsf{\Lambda_r}^{-1}$, which has only r positive (non-zero) diagonal elements, and both U and V are effectively replaced by smaller matrices $\mathsf{U_r}$ and $\mathsf{V_r}$, which have the smaller sets of eigenvectors $\mathsf{u_1 \ldots u_r}$ and $\mathsf{v_1 \ldots v_r}$ $(r \leq m)$, respectively.

Hence the solution $\mathsf{Y} = \mathsf{y} + \delta\mathsf{y}$ is given by

$$\mathsf{Y} = \mathsf{A}^{-1}\mathsf{B} = \mathsf{V_r}\mathsf{\Lambda_r}^{-1}\mathsf{U_r^T}(\mathsf{b} + \delta\mathsf{b}) \tag{6.95}$$

where $\mathsf{B} = \mathsf{b} + \delta\mathsf{b}$, and where A^{-1} is a pseudo inverse operating only on the subspace spanned by $\mathsf{u_1 \ldots u_r}$ and leading to results in the subspace spanned by $\mathsf{v_1 \ldots v_r}$. The measurement error σ is taken into consideration by discarding all the singular values that are smaller than λ_r. This means that both AY and b are in the r dimensional subspace spanned by $\mathsf{u_1 \ldots u_r}$, while $\delta\mathsf{b}$ is in the complementary $m - r$ dimensional orthogonal subspace. Similarly, Y is in the r dimensional subspace spanned by $\mathsf{v_1 \ldots v_r}$, while $\delta\mathsf{y}$ is in the complementary $m - r$ dimensional orthogonal subspace. Therefore, we have

$$||\mathsf{b} + \delta\mathsf{b}||^2 = ||\mathsf{b}||^2 + ||\delta\mathsf{b}||^2 \tag{6.96}$$

and

$$||\mathsf{AY} - \mathsf{B}||^2 = ||\mathsf{AY} - \mathsf{b}||^2 + ||\delta\mathsf{b}||^2, \tag{6.97}$$

and the l.h.s. is at its minimum when $\mathsf{AY} - \mathsf{b} = 0$. When there is no constraint on Y, a solution for $\mathsf{AY} - \mathsf{b} = 0$ can always be found. However, if some of the elements of Y happen to be negative, and the non-negativity constraint is imposed, then Y is no longer at the calculus minimum of $||\mathsf{AY} - \mathsf{b}||^2$ and $||\mathsf{AY} - \mathsf{b}||^2 \geq 0$ at the minimum.

Chapter 7

Theory of NMR in fluid-saturated porous media

Following the earlier work on NMR relaxation of fluid in a confined cell or pore space by Wayne and Cotts [223] and Robertson [187], the experimental findings by Callaghan et al. [50, 51, 52] on the diffraction-like behavior for the magnetization of fluid-saturated porous media, and numerous experimental investigations of the NMR properties of fluid-saturated rocks, there have been many theoretical discussions about the nuclear spin relaxation and diffusion in restricted geometries (e.g. Refs. [45, 95, 165, 43, 18]). In this chapter, we attempt to present a coherent treatment of the subject in a rigorous manner. Some of the mathematical derivations are deferred to the Appendices in order to allow for a clearer presentation in the main text.

7.1 Spin relaxation and diffusion

Spatial diffusion of the nuclei that produce an NMR spin echo can have a crucial effect on the properties of these echoes: When the spin polarizing magnetic field depends on position, the Larmor precession frequency of the nuclear spin of a single diffusing atom keeps changing with time in an incoherent fashion. This leads to an enhancement of the transverse decay rate in a Hahn's spin echo experiment, performed on a uniform or unconstrained fluid, from $1/T_2$ to

$$\frac{1}{T_2} + D_p(\gamma \nabla H)^2 \frac{\tau^2}{3}. \tag{7.1}$$

Here D_p is the diffusion constant, γ is the gyromagnetic ratio of the diffusing spin, ∇H is the local gradient of the polarizing field H, and τ is the time between the initial $\pi/2$ pulse and the π spin inversion pulse, and also between the latter pulse and the spin echo [201]. This expression was derived and explained in Section 2.5.1.

When the fluid in which the precessing spin diffuses is confined either in a small finite volume, or in a connected network of pores of a porous medium, further changes in the transverse relaxation rate occur. The diffusion problem now has many eigenmodes. Each mode decays exponentially in time with its own characteristic decay rate, which is an eigenvalue of the time dependent diffusion problem.

These eigenvalues are also affected by the presence of sinks in the fluid or at its boundaries, where the polarization of a diffusing particle can be destroyed. A uniform volume distribution of such sinks provides a constant extra additive contribution to the intrinsic decay rate $1/T_2$, and its effects can therefore be taken into account by using an altered value for T_2.

More subtle are the effects of a surface distribution of such sinks at the boundaries of the fluid, i.e., at the interface between the pore space and the solid rock matrix. This occurs when there are paramagnetic ions at that interface. The effects of such a "surface absorption" of polarization density are described by a coefficient of surface absorption[1] ρ, whose dimensions are length/time. Obviously, the importance of ρ will depend on the (linear) pore size a, since the absorption at the interface can only occur if the diffusing particle has reached it.

The competition between diffusion and surface absorption leads to the concept of fast versus slow diffusion: The diffusion is *fast* if $\rho a/D_p \ll 1$, and is *slow* if $\rho a/D_p \gg 1$. Clearly, in a uniform fluid, one is always in the slow diffusion limit, since $a \to \infty$. One might therefore think that, in this limit, the diffusion will be less sensitive to the structure of the pore space. However, we shall see in Section 7.1.1 below that this conclusion is premature, and that the slow diffusion limit in fact has the potential, under certain conditions, of providing detailed information about the pore structure by studying the effects of diffusion on NMR spin echoes.

In order to estimate the magnitude of $\rho a/D_p$ in typical porous rocks, we

[1] This constant is referred to as "surface relaxivity" in earlier chapters. Here, we use a more general term "surface absorption" in order to broaden the scope of applicability of the solution for the diffusion problem, which is useful in many other areas such as catalysis, gas absorption, etc.

note that ρ in sandstones is usually in the range 1–10 μm/sec (see Section 4.1.2, Table 4.1, and Ref. [32]), while the mean pore diameter a is in the range 0.0015–0.015 cm [118].[2] Using the value $D_p = 2.5 \times 10^{-5}$ cm^2/sec for the self diffusion coefficient of pure water, we get

$$\frac{\rho a}{D_p} = 0.006 - 0.6.$$

This range is almost entirely within the fast diffusion regime. Nevertheless, we will consider the entire range of possible values of the dimensionless parameter $\rho a/D_p$, including both slow and fast diffusion regimes. This is done in order to show how measurements of spin echoes can be used to obtain useful information about the effects of pore structure on the macroscopic diffusion, and subsequently also on the permeability to flow of brine or hydrocarbon fluids through that structure.

7.1.1 Diffusion in restricted geometries with a uniform polarizing field

When there is both a volume distribution ν and a surface distribution ρ of spin absorbing sinks, the spin density $m(\mathbf{r}, t)$ satisfies the following equations

$$\begin{aligned} \frac{\partial m}{\partial t} &= \nabla \cdot (D_p \nabla m) - \nu m \quad \text{in the pore space,} && (7.2)\\ 0 &= \mathbf{n} \cdot D_p \nabla m + \rho m \quad \text{at the pore/matrix interface,} && (7.3)\\ m(\mathbf{r}, t=0) &= \frac{M(0)}{V_p} \quad \text{at the initial time.} && (7.4) \end{aligned}$$

Here $\mathbf{n}$ is the outward pointing unit vector normal to the pore/matrix interface (i.e., it points out of the pore space and into the solid rock), while $M(0)/V_p$ is the initial spin density, assumed to be uniform in the pore space of volume V_p. We note that D_p was not taken outside (i.e., to the left of) the $\nabla\cdot$ operator in the differential equation: That will allow us to use these equations to describe a situation where D_p is not a constant, but varies with position. This is sometimes a useful option, as explained in Appendix A.

In general, the spin density is actually a complex vector quantity, which represents a non-equilibrium spin polarization produced by a uniform, resonant rf field which is applied to the system for a short period of time before

[2] Note that the values quoted for ρ in this reference are too large, by more than an order of magnitude, as shown by the more recent measurements described in Chapter 4.

$t = 0$. E.g., after a π pulse of the rf field, $M(0)/V_p$ is a real vector opposite to the fixed polarizing field H_0, while after a $\pi/2$ pulse, $M(0)/V_p$ represents a complex vector, perpendicular to H_0. Because the direction and complex phase of $m(\mathbf{r}, t)$ are not altered by the diffusive motion, we ignore those features when solving the equations for $m(\mathbf{r}, t)$.

Note that the diffusion problem is set up with only two parameters, ν and ρ. Thus the solution is applicable for either T_1 or intrinsic T_2, where ν or ρ would have a subscript 1 or 2, depending on whether m represents a polarization that is parallel or transverse to the fixed uniform spin polarizing magnetic field.

However, the function $m(\mathbf{r}, t)$ will be unable to describe the enhanced relaxation rate of transverse polarization due to field gradients, as in Eq. (7.1), because we did not include the field gradient contribution $-i\gamma(\mathbf{r}\cdot\nabla H)m$ to the rate of change of $m(\mathbf{r}, t)$ in Eq. (7.2), as we did in the Torrey equation (2.25) of Section 2.5.1. The field gradient term leads to position and time dependent changes in the complex phase of $m(\mathbf{r}, t)$. The Torrey equation was solved there for the case of a uniform fluid, where that term was found to enhance the transverse relaxation rate, as described by Eq. (7.1). That treatment of the field gradient is only possible in the case of an infinite uniform fluid.

In order to consider the dephasing and the refocusing of spins in a T_2 measurement in *restricted geometries*, we need to use the diffusion propagator which will be introduced in Section 7.1.2. For now, we will confine ourselves to a discussion of intrinsic T_2 due to volume or surface sinks only, and assume that the field is homogeneous and that the phases of all spins are identical.

Diffusion eigenstates

The eigenstates $m_n(\mathbf{r}, t)$ of this problem are states where the spatial density maintains a fixed shape and only its overall magnitude decays exponentially with time, i.e.

$$m_n(\mathbf{r}, t) = \psi_n(\mathbf{r})e^{-\lambda_n t}.$$

The eigenfunctions $\psi_n(\mathbf{r})$ and eigenvalues λ_n satisfy the equations

$$-\nabla \cdot (D_p \nabla \psi_n) + \nu\psi_n = \lambda_n \psi_n \quad \text{in the pore space,} \qquad (7.5)$$

$$\mathbf{n} \cdot D_p \nabla \psi_n + \rho\psi_n = 0 \quad \text{at the pore/matrix interface.} \qquad (7.6)$$

The time dependent spin density or polarization density $m(\mathbf{r}, t)$ can be expanded in terms of the eigenstates as

$$m(\mathbf{r},t) = \sum_{n=0}^{\infty} A_n \psi_n(\mathbf{r}) e^{-\lambda_n t}. \tag{7.7}$$

Since the eigenfunctions are mutually orthogonal, the expansion coefficients A_n can be found from the initial uniform density $m(\mathbf{r}, t=0) = M(0)/V_p$ by (all the integrals here are over the volume V_p of the pore space)

$$A_n = \frac{M(0)}{V_p} \frac{\int dV \psi_n^*(\mathbf{r})}{\int dV |\psi_n(\mathbf{r})|^2}.$$

The total time dependent polarization is then given by

$$M(t) = \int dV m(\mathbf{r},t) = M(0) \sum_{n=0}^{\infty} f_n e^{-\lambda_n t}, \tag{7.8}$$

where the relative intensities or absolute probabilities f_n are given by

$$f_n = \frac{1}{V_p} \frac{\left|\int dV \psi_n(\mathbf{r})\right|^2}{\int dV |\psi_n(\mathbf{r})|^2}, \tag{7.9}$$

and sum up to 1

$$\sum_{n=0}^{\infty} f_n = 1. \tag{7.10}$$

Infinite uniform fluid

In a uniform fluid that fills up all space, the lowest eigenvalue λ_0, corresponding to the longest relaxation time $\tau_0 = 1/\lambda_0$, is equal to ν, since $\psi_0(\mathbf{r}) \equiv$ const. Therefore ν can always be taken as the bulk transverse relaxation rate $1/T_{2B}$ plus any additional relaxation rate ν_0 that is produced by a volume distribution of sinks. In the following discussion, we shall simply use ν to include all these bulk effects, i.e., $\nu = 1/T_{2B} + \nu_0$, because they do not play any role in the special relaxation properties of porous media. Each of the other eigenstates can be completely characterized by a wave vector $\mathbf{q}$, which can take any value. The eigenfunctions are $e^{i\mathbf{q}\cdot\mathbf{r}}$, and the eigenvalues are

$$\lambda_{\mathbf{q}} = \nu + D_p \mathbf{q}^2, \tag{7.11}$$

which is obtained by substituting $e^{i\mathbf{q}\cdot\mathbf{r}}$ for ψ_n in Eq. (7.5). Here, and in every other case where ν has a constant value, we can always ignore ν when calculating the eigenvalues λ_n, and add it to those eigenvalues at the end of the calculation.

In the case of a uniform fluid, the continuous spectrum of relaxation rates of Eq. (7.11) makes no contribution to the relaxation rate of the total polarization $M(t)$: The only nonzero expansion coefficients in Eqs. (7.7) and (7.8) are A_0, f_0, since all the other eigenfunctions are orthogonal to ψ_0, which is constant, hence their integrals vanish. Therefore

$$\frac{M(t)}{M(0)} = e^{-\lambda_0 t} = e^{-\nu t}. \tag{7.12}$$

Restricted geometries

In restricted geometries, this is no longer true. These eigenfunctions, ψ_n, reflect the characteristics of the confined geometries. However, depending on whether the system is in the fast or slow diffusion regime, the information about the microstructure of the pore space may not be extractable from measurements of $M(t)$.

Fast diffusion limit In the *fast diffusion limit* $\rho a/D_p \ll 1$, the boundary condition at the pore/matrix interface reduces to $\mathbf{n}\cdot\nabla m \cong 0$. Consequently $\psi_0(\mathbf{r}) \cong$ const and $f_0 \cong 1$ [from Eq. (7.9)], so that the relaxation of $M(t)$ is again a one exponential decay characterized by the slowest decay rate λ_0, which can be found, approximately, by the following calculation:

$$(\lambda_0 - \nu)\int dV\psi_0 = -\int dV\nabla\cdot(D_p\nabla\psi_0) = -\oint dSD_p\tfrac{\partial\psi_0}{\partial n} = \oint dS\rho\psi_0$$
$$\Rightarrow \lambda_0 \cong \nu + \tfrac{\oint dS\rho}{V_p} \equiv \nu + \overline{\rho}\tfrac{S_p}{V_p}, \tag{7.13}$$

where we used Eqs. (7.5) and (7.6). The volume integrals here are over the pore space of volume V_p, while the surface integrals $\oint dS$ are over the pore/matrix interface of area S_p; $\overline{\rho}$ is the average of ρ over the interface, and S_p/V_p is the surface-to-volume ratio of the pore space, which is a good quantitative measure for the inverse of the typical linear size a of that space: It is valid for isolated pores of finite volume, as well as for a connected pore space that extends to unlimited distances.

However, S_p/V_p contains *no information* at all about the *shape* of the pore space, in particular, it does not provide any information about the existence of bottlenecks in that space, such as narrow necks or narrow cracks which separate between larger pores and can drastically impede both diffusion and fluid flow. In the case of isolated pores with simple shapes, it is easy to write an explicit form for this ratio, e.g., for a sphere of radius a we have $S_p/V_p = 3/a$, while for a circular cylinder of radius a we have $S_p/V_p = 2/a$, and for a parallel-slab-shaped pore of thickness $2a$ we have $S_p/V_p = 1/a$.

Because $\psi_0(\mathbf{r})$ is almost independent of $\mathbf{r}$ inside the pore space, the probability f_0 is very close to 1. Therefore all other eigenstates, with decay rates greater than λ_0, make a negligible contribution to $M(t)$. Consequently $M(t)$ exhibits simple exponential decay at a rate given by Eq. (7.13).

Slow diffusion limit For isolated pores with simple shapes (i.e., sphere, infinitely long circular cylinder, flat region between two parallel planes), the eigenfunctions and eigenvalues can be calculated quite explicitly, in terms of trigonometric or Bessel functions and their zeros [45]. From these solutions we learn that, in the *slow diffusion limit* $\rho a/D_p \gg 1$ and for fixed pore shape, all eigenvalues satisfy $\lambda_n = g_n^2 D_p/a^2$. The eigenfunctions are *universal functions* (i.e., their form depends only on the pore shape) of $g_n\mathbf{r}/a$, where the dimensionless constants g_n are of order 1 and depend on the eigenvalue index n. They also depend on the pore shapes in a non-trivial fashion. Therefore, even if D_p is known, a measurement of just one relaxation time is unable to provide detailed information about the size and shape of the pore space. A fortiori, interesting physical properties, like fluid permeability, will not be determined by such a measurement. Only a measurement of the *distribution of relaxation rates* of $M(t)$, i.e., of both λ_n and f_n for many different values of n, has a chance of providing information about the topology, and consequently about the fluid permeability, of the porous system. In the slow diffusion regime, many decay modes contribute significantly to the decay of $M(t)$. Hence a non-trivial distribution of relaxation rates can be obtained even by measurements of limited accuracy.

It should be stressed that, although the slow diffusion limit has a greater potential for providing detailed microstructural information, it is not desirable to have D_p/a^2 too small: If that quantity is much smaller than the relaxation rate due to volume effects, like the pure fluid relaxation rate $1/T_{2B}$ and the volume distribution of sinks ν_0, then it will be difficult to resolve the spectrum of microstructure dependent relaxation rates λ_n. Ideally, we would like to have

$$\frac{1}{T_{2B}} + \nu_0 < \frac{D_p}{a^2} \text{ and } \frac{\rho a}{D_p} > 1,$$

which entails

$$\frac{\rho}{a} > \frac{1}{T_{2B}} + \nu_0. \tag{7.14}$$

Using the ranges quoted above for ρ and a (see Section 7.1), we find

$$\frac{\rho}{a} = 0.007 - 0.7 \text{ sec}^{-1}$$

for the typical range of values of their ratio. Since $T_{2B} < T_{1B} \cong 1$ sec for the relaxation time of protons in pure water, this means that the inequality of Eq. (7.14) is not satisfied. Therefore measurements of the distribution of relaxation rates in a uniform polarizing field will not be useful as a means to obtain detailed information about microstructure of the pore space.

Initial rate of decay Another interesting quantity that can sometimes be measured (see, e.g., Section 3.2.2) is the initial slope of the decaying total polarization $\dot{M}(0)$, also denoted by $-M(0)/T_{in}$, where $1/T_{in}$ is the initial relaxation rate. Using Eqs. (7.2) and (7.3), we can write the following expression for the instantaneous decay rate of $M(t)$:

$$\begin{aligned} -\dot{M}(t) &= -\int dV \dot{m}(\mathbf{r}, t) = -\int dV [\nabla \cdot (D_p \nabla m) - \nu m] \\ &= -\oint dS D_p \mathbf{n} \cdot \nabla m + \int dV \nu m = \oint dS \rho m + \int dV \nu m. \end{aligned}$$

At the initial time $t = 0$, we have $m(\mathbf{r}, t = 0) = M(0)/V_p$ in the pore space, therefore

$$\frac{1}{T_{in}} \equiv -\frac{\dot{M}(0)}{M(0)} = \nu + \overline{\rho} \frac{S_p}{V_p}. \tag{7.15}$$

This is the same as λ_0 in the fast diffusion limit—see Eq. (7.13). In that limit, the initial decay rate of $M(t)$ persists for all times, whereas for the slow diffusion limit, it is only true for the initial decay.

Summary In the slow diffusion limit, and also in the intermediate regime between the fast and the slow diffusion regimes, there is a non-trivial distribution of relaxation times for $M(t)$. These relaxation times depend on the pore sizes and shapes in a way that also depends on the surface absorption coefficient ρ. Therefore, if both D_p and ρ are known, measurements of this distribution in these regimes can provide detailed information about the topology of the pore space, and consequently about the fluid flow permeability. By contrast, in the fast diffusion limit the decay of $M(t)$ is a simple exponential function at all times, with a fixed decay rate equal to the initial rate of decay. That rate contains no information about the shape of the pore space, other than the ratio S_p/V_p.

7.1.2 The diffusion propagator

Before considering the effect of diffusion of spins in a fluid-saturated porous medium where the magnetic field is non-uniform, it is useful to introduce the concept of "diffusion propagator" $G(\mathbf{r},\mathbf{r}'|t) = G(\mathbf{r}',\mathbf{r}|t)$. This is the conditional probability density to find a particle at $\mathbf{r}'$ at time $t > 0$, if a particle was present at $\mathbf{r}$ at time $t = 0$. In fact, this is just the Green's function for the diffusion problem in the pore space, and it satisfies the following equations (here we assume that D_p is constant throughout the pore space)

$$\frac{\partial G}{\partial t} = D_p\nabla^2 G - \nu G \quad \text{in the pore space,} \tag{7.16}$$

$$0 = \frac{\partial G}{\partial n} + \frac{\rho}{D_p}G \quad \text{at the pore/matrix interface,} \tag{7.17}$$

$$G(\mathbf{r},\mathbf{r}'|t=0) = \delta^3(\mathbf{r}-\mathbf{r}'). \tag{7.18}$$

In Section 7.2 below it is shown that the average of this propagator can be found by measuring an NMR spin echo in a "pulsed-field-gradient-spin-echo (PFGSE)" experiment. This is a powerful experimental technique which we discussed in Chapter 2, and it is frequently used for measuring the microstructure and diffusion properties of porous media: Since the time and spatial dependence of G reflect the detailed microstructure of the pore space, such measurements can provide rather detailed information about that microstructure. Even the more conventional measurements of NMR spin echos in the presence of a fixed gradient of the polarizing field (i.e. "fixed-gradient-spin-echo" or FGSE), provide some information about the microstructure and the diffusion in such media.

The PFGSE amplitude $M(\mathbf{k},t)$, which is measured in a PFGSE experiment, is related to the diffusion propagator by [see Eq. (7.61) below]

$$M(\mathbf{k},t) \equiv \frac{1}{V_p}\int_{V_p} dV \int_{V_p} dV' G(\mathbf{r},\mathbf{r}'|t) e^{-i\mathbf{k}\cdot(\mathbf{r}-\mathbf{r}')}. \quad (7.19)$$

Here t is the time separation between the field gradient pulses, V_p is the total volume of the pore space, and the wave vector $\mathbf{k}$ is simply related to the polarizing field gradient ∇H, the length of time δ during which the two field gradient pulses are applied, and the gyromagnetic ratio γ of the polarized spin

$$\mathbf{k} = \gamma\delta\nabla H. \quad (7.20)$$

The PFGSE experimental method was described in Section 2.5.2. More details about the specific experimental procedure for measuring $M(\mathbf{k},t)$ and the significance of the various parameters are provided in Section 7.2 below. The microstructure of the pore space is assumed to be "macroscopically homogeneous and isotropic", therefore the average of $G(\mathbf{r},\mathbf{r}'|t)$ over different realizations of that microstructure (i.e., the ensemble average) will depend only on $|\mathbf{r}-\mathbf{r}'|$. In that case, $M(\mathbf{k},t)$ includes all the information needed to reproduce that average G. Also, the single volume integral

$$\int_{V_p} dV' G(\mathbf{r},\mathbf{r}'|t),$$

as well as $M(\mathbf{k},t)$, are dimensionless and of order 1.

Upon close examination of Eq. (7.19), it is not too difficult to realize that the PFGSE amplitude is simply the average, over all spins, of the phase factor acquired when the Larmor frequency was momentarily changed by turning on a uniform field gradient ∇H for two short periods of time, of length δ each and separated by a time interval of length t, before and after application of the spin flipping π pulse of rf radiation (also, see the discussion in Section 7.2 below). This is different from the regular Hahn spin echo experiment, in which the spins are always under the influence of a time independent field gradient, i.e., a "fixed-gradient-spin-echo (FGSE)". However, we shall see later that many important quantities of the porous media, including the FGSE, can be derived using the diffusion propagator and the PFGSE amplitude. Thus, both of them are very useful in the theoretical treatment of spin diffusion in porous media.

The magnetization or spin density $m(\mathbf{r}, t)$ inside the pore space, given that $m(\mathbf{r}, 0) = M(0)/V_p$, can be found from

$$m(\mathbf{r}, t) = \frac{M(0)}{V_p} \int_{V_p} dV' G(\mathbf{r}, \mathbf{r}'|t). \tag{7.21}$$

The spatial integral of this quantity is just the total polarization $M(t)$, therefore

$$\frac{M(t)}{M(0)} = \frac{1}{V_p} \int_{V_p} dV \int_{V_p} dV' G(\mathbf{r}, \mathbf{r}'|t) = M(\mathbf{k} = 0, t). \tag{7.22}$$

This is just the special case of $M(\mathbf{k}, t)$ when $\mathbf{k} = 0$, i.e., for the uniform field case when $\nabla H = 0$.

If there is no absorption or decay of spin polarization, i.e. $\rho = \nu = 0$, then, using Eqs. (7.16) and (7.17), we get:

$$\begin{aligned} \frac{d}{dt} \int_{V_p} dV G(\mathbf{r}, \mathbf{r}'|t) &= \int_{V_p} dV D_p \nabla^2 G = \oint_{\partial V_p} dS\, D_p \frac{\partial G}{\partial n} = 0 \\ &\Rightarrow \dot{M}(t) \equiv 0. \end{aligned}$$

In this case the total probability for finding the diffusing spin somewhere in the pore space is constant in time and equal to 1, and $M(t) \equiv M(0)$.

By contrast, when ρ and/or ν are nonzero, that probability decreases monotonically to 0 with increasing t, and the total spin polarization also diminishes gradually as a result of the surface and volume sinks. A nonzero constant value of ν merely contributes the exponential factor $e^{-\nu t}$ to G, and this factor is independent of microstructure. Therefore it can be left out of all microstructure dependent calculations and only be brought in at the end as a multiplicative factor. By contrast, a nonzero value of ρ, even if it is constant, will contribute to both the time dependence and the spatial form of G.

Infinite uniform fluid

In a uniform pure fluid, the diffusion propagator has the well known Gaussian form

$$G_0(\mathbf{r}, \mathbf{r}'|t) = \frac{1}{(4\pi D_p t)^{3/2}} \exp\left[-\frac{(\mathbf{r} - \mathbf{r}')^2}{4D_p t} - \nu t\right]. \tag{7.23}$$

This can easily be verified by substitution in the diffusion equation (7.2) or (7.16). The subscript 0 is used to identify quantities related to the infinite uniform fluid.

Recalling the probabilistic significance of G_0, it is easy to calculate the mean square distance travelled by a particle (or "mean square displacement"), diffusing in a uniform fluid, during time t

$$\langle[\mathbf{r}(t)-\mathbf{r}(0)]^2\rangle = \frac{\int dV'(\mathbf{r}-\mathbf{r}')^2 G_0(\mathbf{r},\mathbf{r}'|t)}{\int dV' G_0(\mathbf{r},\mathbf{r}'|t)} = 6D_p t. \tag{7.24}$$

Note that this result is independent of the existence of any uniform intrinsic relaxation rate $1/T_2$ or uniform volume distribution of sinks ν. That is so because particles that get absorbed during the time t do not contribute to the mean square displacement. In such a system, the PFGSE amplitude can be obtained by substituting Eq. (7.23) into Eq. (7.19):

$$M_0(\mathbf{k},t) = e^{-D_p \mathbf{k}^2 t - \nu t}. \tag{7.25}$$

Setting $\mathbf{k}=0$ leads to

$$\frac{M_0(t)}{M(0)} = e^{-\nu t},$$

which is the same as Eq. (7.12).

Restricted geometries

For restricted geometries, the diffusion propagator no longer has a simple Gaussian form as in Eq. (7.23). It can always be expanded using the diffusion eigenstates:

$$G(\mathbf{r},\mathbf{r}'|t) = \frac{1}{V}\sum_{\lambda} e^{-\lambda t}\psi_\lambda(\mathbf{r})\psi_\lambda^*(\mathbf{r}'), \tag{7.26}$$

where V is the total volume of the system, including both pore space and rock matrix. The eigenvalues λ and eigenfunctions $\psi_\lambda(\mathbf{r})$ are obtained by solving the following eigenvalue problem

$$\lambda\psi_\lambda + D_p\nabla^2\psi_\lambda = 0 \quad \text{inside the pore space,} \tag{7.27}$$

$$D_p\frac{\partial\psi_\lambda}{\partial n} + \rho\psi_\lambda = 0 \quad \text{at the pore/matrix interface.} \tag{7.28}$$

These equations are the same as Eqs. (7.5) and (7.6), but the eigenfunctions are now normalized, in addition to being mutually orthogonal,

$$\frac{1}{V}\int_{V_p} dV \psi_\lambda^* \psi_\nu = \delta_{\lambda\nu}. \tag{7.29}$$

From Eqs. (7.22) and (7.26), the total spin polarization $M(t)$ can be written as a sum of exponentially decaying terms

$$\frac{M(t)}{M(0)} = \frac{1}{\phi}\sum_\lambda e^{-\lambda t}\left|\frac{1}{V}\int_{V_p} dV\, \psi_\lambda(\mathbf{r})\right|^2, \qquad \phi \equiv \frac{V_p}{V}. \tag{7.30}$$

For a general disordered porous medium, there is no simple way to obtain these eigenstates and hence the diffusion propagator. However, for periodic porous media, it is possible to numerically compute both quantities for the case where $\rho = 0$, as well as for the case where $\rho \neq 0$. This is illustrated in Appendix A, where we describe some calculations of the eigenstates, and of $M(\mathbf{k}, t)$ of Eq. (7.19), i.e., the spatial Fourier transform of the average G, for such media. Using these numerical solutions, many interesting quantities for porous media, such as PFGSE amplitude, enhanced T_2 relaxation in heterogeneous fields, etc., can be calculated, as we shall discuss in subsequent sections.

Although it is not possible to compute the diffusion propagator for a general disordered porous medium, it is possible to obtain its asymptotic behavior in certain limiting cases. From those asymptotic results the polarization decay $M(t)$ can also be calculated.

Short time behavior For short times t, i.e., $D_p t \ll a^2$ and $\rho t \ll a$, where a is a characteristic linear size of the pore space, the changes in the form of $G(\mathbf{r}, \mathbf{r}'|t)$ can be calculated to leading order in the small quantities $D_p t/a^2$ and $\rho t/a$—see Appendix B. For $M(t)/M(0)$ this leads to the following short time asymptotics (we omit the factor $e^{-\nu t}$ here; see Eq. (B.3) in Appendix B and Ref. [167])

$$\frac{M(t)}{M(0)} \cong \begin{cases} 1 - \frac{S_p}{V_p}\rho t & \text{for small } \rho \ll (D_p/t)^{1/2}, \\ 1 - 2\frac{S_p}{V_p}\left(\frac{D_p t}{\pi}\right)^{1/2} & \text{for large } \rho \gg (D_p/t)^{1/2}. \end{cases} \tag{7.31}$$

The first line here, which is always valid for sufficiently small ρ, agrees with results obtained earlier for $1/T_{in}$—see Eq. (7.15). The second line shows that even when ρ is very large, the short time decay of $M(t)$ depends only on the ratio S_p/V_p, and is insensitive to the shape of the pore space. In particular, it does not distinguish between a connected pore space and a collection of isolated voids, if they happen to have the same S_p/V_p ratio.

Long time behavior: The macroscopic stationary diffusion coefficient For sufficiently long times, and for $\mathbf{k}$ that is not too large, the PFGSE amplitude has a form similar to that of the uniform fluid function $M_0(\mathbf{k}, t)$ of Eq. (7.25), but with the free fluid diffusion constant D_p replaced by the "bulk effective (or macroscopic) stationary diffusion coefficient" D_e

$$\frac{M(\mathbf{k}, t)}{M(\mathbf{k}=0, t)} \cong e^{-D_e \mathbf{k}^2 t} \text{ for } D_p t \gg a^2 \text{ and } |\mathbf{k}a| \ll 1. \tag{7.32}$$

Low lying eigenvalues and electrical conductivity The value of D_e is determined by the low lying eigenstates of the diffusion problem. This macroscopic diffusion coefficient is related to the eigenvalue spectrum of the diffusion equation. However, in contrast with the case of the uniform fluid, the eigenfunctions are no longer simple plane waves, and the eigenvalues do not have the simple form of Eq. (7.11).

When $\rho = 0$, the lowest eigenfunction is still a constant, but when $\rho \neq 0$ it is a non-constant function $\psi_0(\mathbf{r})$. All we can say, in general, about $\psi_0(\mathbf{r})$ is that it is real and positive throughout the connected part of the pore space, and that it is *macroscopically uniform*, namely, it does not exhibit any tendency to be localized in some small part of the pore space [21]. The detailed form of $\psi_0(\mathbf{r})$ depends on the shape of the pore space, and the same is true for the other eigenfunctions. However, the *low lying eigenstates* can be obtained, approximately, in a somewhat general fashion once $\psi_0(\mathbf{r})$ is known. Those states are characterized by a wave vector $\mathbf{q}$, as in the case of the uniform fluid, and have the following form when $|\mathbf{q}|a \ll 1$ [21]:

$$\psi_{\mathbf{q}}(\mathbf{r}) = \psi_0(\mathbf{r}) e^{i \sum_\alpha q_\alpha \phi_\alpha(\mathbf{r})} + \mathcal{O}(\mathbf{q}^2), \tag{7.33}$$

$$\lambda_{\mathbf{q}} = \lambda_0 + D_e \mathbf{q}^2 + \mathcal{O}(\mathbf{q}^4). \tag{7.34}$$

Here $\phi_\alpha(\mathbf{r})$ is the electrical potential of a conductivity problem in the pore space, with local conductivity equal to $\psi_0^2(\mathbf{r})\sigma_p$ and an average electric field equal to $\mathbf{e}_\alpha$ (the unit vector along the α axis), and σ_p is the conductivity of the pore fluid. This leads to

$$\frac{D_e}{D_p} = \frac{\int_{V_p} dV \psi_0^2 (\nabla \phi_\alpha)^2}{\int_{V_p} dV \psi_0^2}. \tag{7.35}$$

Here we are assuming, as always, that D_e has the same value in all directions, i.e., that the microstructure is macroscopically isotropic. This expression can easily be generalized to cases where the microstructure is anisotropic, and hence the bulk effective diffusion coefficient D_e becomes a second rank tensor instead of a scalar quantity [21].

The value of D_e when $\rho = 0$ is closely related to the value of the macroscopic or bulk effective electrical conductivity σ_e of the fluid-saturated porous medium:

$$\frac{D_e}{D_p} = \frac{1}{\phi}\frac{\sigma_e}{\sigma_p} = \frac{1}{\phi F}, \tag{7.36}$$

where $F \equiv \sigma_p/\sigma_e$ is the "formation factor". Note that here $\phi \equiv V_p/V$ denotes the volume fraction of the pore space or the porosity, and *not* the electrical potential ϕ_α of Eq. (7.35).

An important property of D_e is that it is always less than D_p [21]. Again, a measurement of $M(\mathbf{k}, t)$ for long times and small $\mathbf{k}$ does not yield any detailed information about the pore space.

When ρ is negligible, $\psi_0(\mathbf{r})$ is constant in the pore space. In that case, σ_e is the *actual electrical conductivity* of the fluid filled porous medium (assuming that the rock matrix does not conduct), which can be measured in standard ways. The measurement of D_e will provide us, however, with an independent method for obtaining the formation factor as a verification of the standard electrical measurement, if ϕ and D_p are known.

It is worth noting that, when $\rho = 0$, D_e/D_p is independent of the absolute size scale a of the pore space, i.e., if the entire microstructure is uniformly expanded or contracted by some constant factor, both D_e and σ_e remain unchanged. This is no longer true when $\rho \neq 0$. In the latter case, D_e does depend on the absolute size scale of the pore space. For this reason, measurements of D_e can then provide more information about the microstructure than when $\rho = 0$.

We note that the low lying eigenvalues of the *porous medium* form a continuous spectrum that looks much like the continuous spectrum of all the eigenvalues of the diffusion problem in a *uniform fluid*—compare Eqs. (7.11) and (7.34). This is why measurement of the asymptotic long time behavior of the macroscopic diffusion can only provide us with the two quantities λ_0 and D_e, and why measurements at finite times are needed to obtain more detailed information about the microstructure.

Time dependent diffusion coefficient Because $G(\mathbf{r}, \mathbf{r}'|t) = G(\mathbf{r}', \mathbf{r}|t)$, therefore $M(-\mathbf{k}, t) = M(\mathbf{k}, t)$. Thus, for small $\mathbf{k}$, $M(\mathbf{k}, t)$ must be a quadratic function of $\mathbf{k}$. One can define a "time dependent diffusivity" $D(t)$, such that [167]:

$$M(\mathbf{k}, t) \cong M(\mathbf{0}, t)e^{-D(t)\mathbf{k}^2 t} \quad \text{for } |\mathbf{k}|a \ll 1. \tag{7.37}$$

The function $D(t)$ starts out from $D(0) = D_p$, and ends up at $D(\infty) = D_e < D_p$, decreasing monotonically with increasing t. Figure 7.1 shows results of calculations of $D(t)$ for a pore space of a periodic simple cubic array of solid spheres. Neighboring spheres in this array can be either non-overlapping, partially overlapping, or just touching. In the case of just touching spheres, the porosity is $\phi = 1 - \pi/6 \cong 0.4764$. Results are shown for a variety of different porosities, ranging from non-overlapping to partially overlapping neighboring spheres, and for a variety of values of ρ.

As in the case of $M(t)$, short time asymptotics can be applied to get [see Eq. (B.4) in Appendix B]

$$\frac{D(t)}{D_p} \cong \begin{cases} 1 - \frac{4}{9}\frac{S_p}{V_p}\left(\frac{D_p t}{\pi}\right)^{1/2} & \text{for } \rho \ll (D_p/t)^{1/2}, \\ 1 - \frac{2}{9}\frac{S_p}{V_p}\left(\frac{D_p t}{\pi}\right)^{1/2} & \text{for } \rho \gg (D_p/t)^{1/2}. \end{cases} \tag{7.38}$$

Interestingly, to the order exhibited here, both results are independent of ρ.

Experiments have been conducted to measure the time dependent diffusion coefficient $D(t)$ as a function of time for water-saturated rocks [108, 139]. The signal usually diminishes before $D(t)$ reaches its asymptotic value $D_e = D_p/F\phi$. Later investigations [148, 149], using both laser-spin-polarized and thermally spin-polarized xenon gas, did manage to reach the asymptotic regime and verified this relation, as well as the short time expression in the first line of Eq. (7.38). Figure 7.2 shows results from those experiments.

While the short time and long time behavior of $D(t)$ can only give us values for S_p/V_p and D_e, its behavior at intermediate times is sensitive to other details of the pore structure. Thus, measurements of $D(t)/D_0$ at intermediate times have the potential of providing further information about those microstructural details. Section 7.1.3 provides an illustration of this idea.

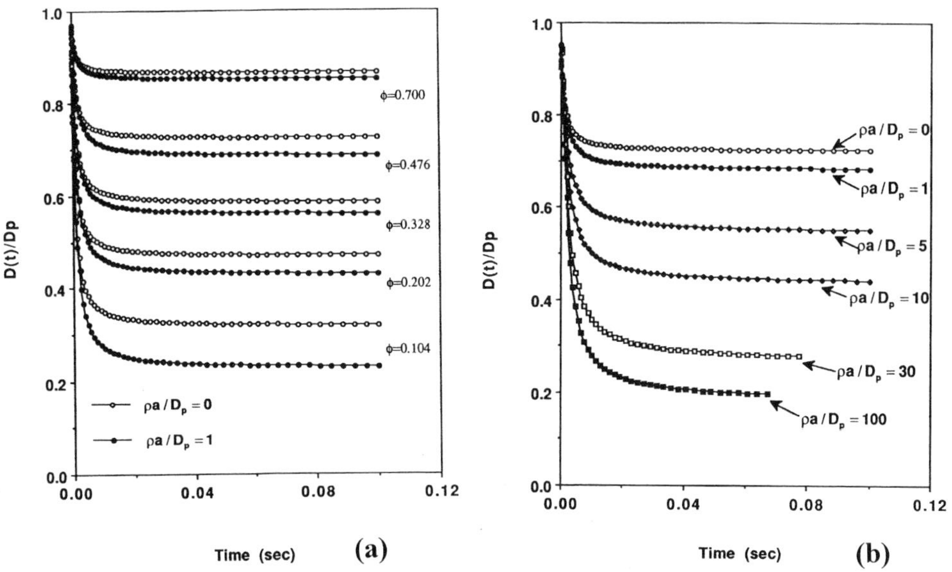

Figure 7.1: Bulk effective time dependent diffusion coefficient $D(t)/D_p$, plotted vs. time t, for a porous medium which is the spatial complement of a periodic, simple cubic array of solid spheres. (a) For various porosities ϕ and two values of the surface absorption coefficient: $\rho = 0$ and $\rho a/D_p = 1$. (b) For touching spheres ($\phi = 0.476$) and various values of $\rho a/D_p$ (after Fig. 3 of Ref. [18]).

7.1.3 Dephasing of the precessing spins

We now discuss the effects of diffusion on the dephasing of a precessing spin in the presence of a nonuniform magnetic field. Such a field can result from a uniform field gradient, intentionally applied in order to achieve some spatial resolution in the spin echo experiment, or it can result from unintentional and uncontrollable random spatial variations of the local magnetic susceptibility. The latter are mostly due to the inherent heterogeneity of the porous rock, i.e., the fact that the magnetic susceptibility χ has different values in the pore fluid and in the rock matrix.

The precession of a spin in a polarizing magnetic field directed along z is conveniently described by the time dependent expectation value $s_+(t)$ of the quantum mechanical spin operator $\hat{s}_x + i\hat{s}_y$. This quantity satisfies a

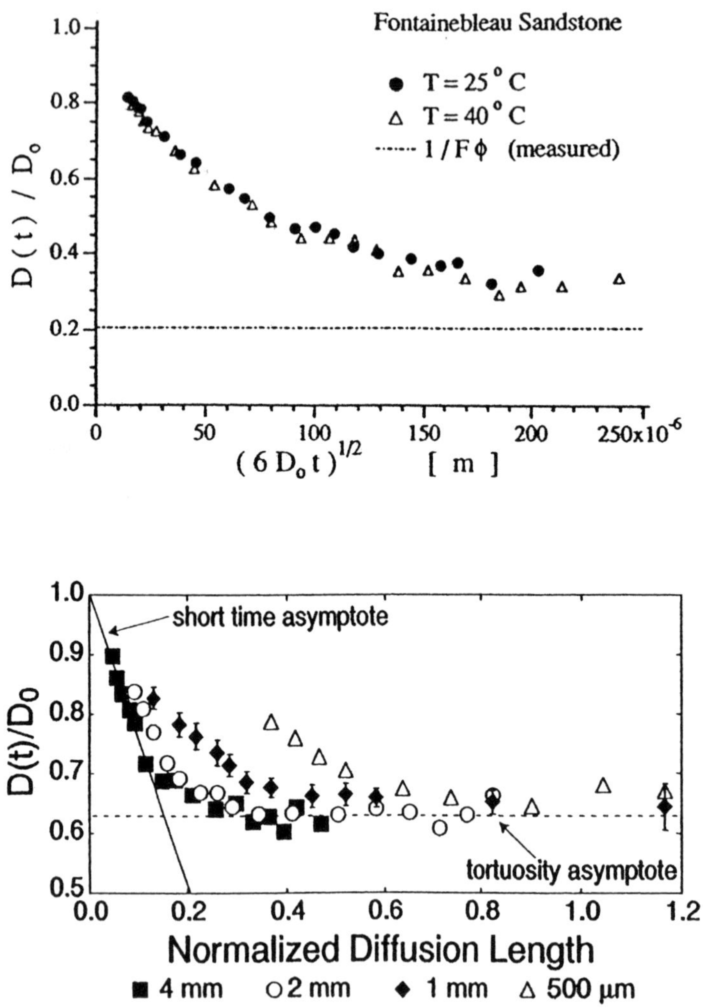

Figure 7.2: The figure on the top shows the bulk effective time dependent diffusion coefficient $D(t)/D_0$, plotted versus $(6D_0t)^{1/2}$, for a Fontainebleau sandstone, where the asymptotic line, $1/F\phi$, was measured independently using electrical method (From Hürlimann et al. [108]). The figure on the bottom shows the bulk effective time dependent diffusion coefficient measurements for thermally polarized xenon gas imbibed in randomly packed spherical glass beads of various sizes, where the asymptotic limit is reached (From Mair et al. [148, 149]).

Langevin type equation of motion

$$\frac{\partial s_+}{\partial t} = -i\gamma\left(H_0 + \delta H[\mathbf{r}(t)]\right) s_+ - \left(\frac{1}{T_{2B}} + \nu_0\right) s_+ + F(t). \tag{7.39}$$

Here γ is the gyromagnetic ratio and we have written the polarizing field as the sum of a fixed uniform field H_0 and a position dependent variation $\delta H[\mathbf{r}(t)]$—the latter contribution depends on the time dependent position $\mathbf{r}(t)$ of the diffusing spin; $1/T_{2B} + \nu_0$ is the transverse relaxation rate, which is a sum of the bulk relaxation rate $1/T_{2B}$ and a possible contribution ν_0 due to a constant volume distribution of sinks; $F(t)$ is a rapidly fluctuating random force term (i.e., noise). Without the latter term, which mimics the microscopic origins of the intrinsic relaxation rate, the solution of this equation is

$$s_+(t) = s_+(0)e^{-i\gamma H_0 t-(1/T_{2B}+\nu_0)t} \exp\left(-i\gamma \int_0^t dt\, \delta H[\mathbf{r}(t)]\right). \tag{7.40}$$

The first exponential factor in this expression includes the rapid precession of the polarized spin, at the Larmor angular frequency $\omega_c = \gamma H_0$ due to the main background polarizing field H_0. It also includes the fixed background relaxation rate $1/T_{2B} + \nu_0$. The second exponential factor contains an extra phase $\Phi(t) = \gamma \int_0^t dt\, \delta H[\mathbf{r}(t)]$, accumulated by the precessing spin due to the position dependent variations in the polarizing field, which cause similar variations in the local Larmor frequency. The fact that this phase is different for different spins is responsible for an enhancement in the magnetization decay for the whole system.

Spin echo A spin echo experiment is best described in a reference frame that rotates around H_0 at the Larmor angular frequency $\omega_c = \gamma H_0$. In that frame, $s_+(t)$ is stripped of this rapid fixed rate of precession and becomes

$$s_{R+}(t) \equiv s_+(t)e^{i\omega_c t} = s_+(0)e^{-(1/T_{2B}+\nu_0)t} \exp\left(-i\gamma \int_0^t dt\, \delta H[\mathbf{r}(t)]\right). \tag{7.41}$$

In order to produce a spin echo at time 2τ, a short π pulse of a resonant rf field is applied at time τ after the initial $\pi/2$ pulse, in the xy plane. The π pulse changes $s_{R+}(t)$ into its complex conjugate. At time 2τ, the total extra phase shift, produced by δH, becomes

$$\Phi(2\tau) = -\gamma \int_0^\tau dt\, \delta H[\mathbf{r}(t)] + \gamma \int_\tau^{2\tau} dt\, \delta H[\mathbf{r}(t)]. \tag{7.42}$$

The sign difference between the phase accumulated before and after the π pulse is due to the complex conjugation of $s_{R+}(\tau)$ caused by that pulse. If the position of the spin $\mathbf{r}(t)$ does not change during the time interval $[0, 2\tau]$, then the two integrals exactly cancel each other and $\Phi(2\tau) = 0$: The dephasing is totally cancelled, and that revives the previously decayed rf radiation at frequency ω_c, and forms the spin echo. The amplitude of that echo is attenuated not only due to the uniform decay rate $1/T_{2B} + \nu_0$, but also due to the interface absorption [see Eq. (7.31)] and the diffusion [see Eq. (7.32)]. In this connection, recall that $M(\mathbf{k}, t)$ represents the amplitude of the spin echo in a PFGSE experiment.

In general, the time dependence of that decay is not well described by a simple exponential, except at very long times and small $\mathbf{k}$, or in the fast diffusion limit. Nevertheless, we sometimes refer to the instantaneous decay rate of the spin echo amplitude *in the absence of any magnetic field heterogeneity effects* as "intrinsic $1/T_2$".

Effects of diffusion using the propagator Because of the diffusive motion, the two terms in Eq. (7.42) do not cancel each other when the field is heterogeneous. Moreover, the degree of non-cancellation differs for spins that move along different paths $\mathbf{r}(t)$. In order to calculate the effects due to this motion, we need to evaluate the average of $e^{-i\Phi(2\tau)}$ over all the spins in the system.

The presence of interface absorption affects the spin echo amplitude in two ways: (a) Even in the absence of any field heterogeneity, $M(t)$ is decreased when $\rho \neq 0$, therefore any spin echo will be similarly attenuated. Eq. (7.31) shows this effect in the limit of very short times. In general, this attenuation is not described by a simple exponential—see Eq. (7.30). (b) A nonzero value of ρ affects the diffusion propagator, and through it the average of $e^{-i\Phi(2\tau)}$. For example, $D(t)$ decreases with increasing ρ—see Fig. 7.1.

In the rotating reference frame, the field heterogeneity is always a small perturbation. Therefore the effect (a) above can be calculated by ignoring that heterogeneity, while the effect (b) can be calculated using the diffusion propagator, which is defined for the pore space *independently of any magnetic field.* The microstructure of the pore space, as well as the interface absorption, enter in the calculation of both effects.

Alternative approach An alternative approach to this calculation can start from Torrey's partial differential equation (2.25). That equation would then have to be solved with the boundary condition of Eq. (7.3) at the pore/matrix interface—a very difficult task. The approach described both above and below avoids that difficulty by treating the field heterogeneity as a *small perturbation.*

Cumulant expansion

One way to calculate the average $\langle e^{-i\Phi(2\tau)}\rangle$ is by expanding its natural (base e) logarithm in a series of cumulant averages $\langle\,\rangle_c$ of powers of $\Phi(2\tau)$, which will henceforth be abbreviated as Φ:

$$\ln\langle e^{-i\Phi}\rangle \equiv \langle e^{-i\Phi} - 1\rangle_c = -i\langle\Phi\rangle_c - \frac{1}{2}\langle\Phi^2\rangle_c + \frac{i}{6}\langle\Phi^3\rangle_c + \frac{1}{24}\langle\Phi^4\rangle_c + \cdots$$

These cumulant averages can be expressed in terms of simple averages $\langle\,\rangle$ (see, e.g., Ref. [147]):

$$\begin{aligned}
\langle\Phi\rangle_c &= \langle\Phi\rangle, \\
\langle\Phi^2\rangle_c &= \langle\Phi^2\rangle - \langle\Phi\rangle^2 \equiv \langle(\Delta\Phi)^2\rangle, \quad \Delta\Phi \equiv \Phi - \langle\Phi\rangle, \\
\langle\Phi^3\rangle_c &= \langle(\Delta\Phi)^3\rangle, \\
\langle\Phi^4\rangle_c &= \langle(\Delta\Phi)^4\rangle - 3\langle(\Delta\Phi)^2\rangle^2, \quad \text{etc.}
\end{aligned}$$

When the distribution of values of Φ is Gaussian, then all cumulants beyond $\langle\Phi^2\rangle_c$ vanish. For non-Gaussian distributions, the magnitudes of $\langle\Phi^3\rangle_c$, $\langle\Phi^4\rangle_c$, etc. are a measure of the deviation from Gaussian shape. The cumulant expansion is useful mainly if those deviations are small.

Averages over possible paths $\mathbf{r}(t)$ of a diffusing particle can be calculated by using the diffusion propagator $G(\mathbf{r}, \mathbf{r}'|t)$. Detailed calculations of such averages have been performed only on periodic microstructures and when $\rho = 0$. The odd order cumulants vanish when $\rho = 0$, including the first cumulant $\langle\Phi\rangle_c = \langle\Phi\rangle$, because time reversed paths have identical probabilities [19]. It follows that the distribution of Φ is approximately Gaussian, and is characterized by the variance of Φ, which is equal to its mean square $\langle\Phi^2\rangle$ when $\langle\Phi\rangle = 0$, if we can show that all higher order cumulant averages are small compared to $\langle\Phi^2\rangle$.

Periodic microstructures In order to illustrate how this is done, we note that in a periodic microstructure, the eigenstates $\psi_\lambda(\mathbf{r})$ have a Bloch-Floquet form, namely,

$$\psi_\lambda(\mathbf{r}) \rightarrow \phi_{n\mathbf{q}}(\mathbf{r})e^{i\mathbf{q}\cdot\mathbf{r}}, \tag{7.43}$$

$$\lambda \rightarrow \lambda_{n\mathbf{q}}, \tag{7.44}$$

where $\phi_{n\mathbf{q}}(\mathbf{r})$ has the same periodicity as the microstructure and $\mathbf{q}$ is a wave vector in the first Brillouin zone of the reciprocal space. The subscript n characterizes the different bands of eigenvalues $\lambda_{n\mathbf{q}}$, which are continuous functions of the $\mathbf{q}$-vector. We have used here, for the first time, the notation $\phi_{n\mathbf{q}}$, $\lambda_{n\mathbf{q}}$ to distinguish the $\rho = 0$ eigenfunctions and eigenvalues from the more general quantities $\psi_{n\mathbf{q}}$, $\mu_{n\mathbf{q}}$ of a periodic porous medium with arbitrary, nonzero values of ρ.

The eigenvalues of the lowest band have the form

$$\lambda_{0\mathbf{q}} = \lambda_{00} + D_e\mathbf{q}^2 + \mathcal{O}(\mathbf{q}^4) \quad \text{for} \quad |\mathbf{q}|a \ll 1. \tag{7.45}$$

The diffusion propagator now has the following form

$$G(\mathbf{r}, \mathbf{r}'|t) = \frac{1}{V}\sum_{n\mathbf{q}} e^{-\lambda_{n\mathbf{q}}|t|}\phi_{n\mathbf{q}}(\mathbf{r})\phi^*_{n\mathbf{q}}(\mathbf{r}')e^{i\mathbf{q}\cdot(\mathbf{r}-\mathbf{r}')}, \tag{7.46}$$

where $\lambda_{n\mathbf{q}}$ are the eigenvalues of the n-th band and V is the total volume of the pore space and matrix.

To compute $\langle\Phi^2\rangle$, we note that

$$\begin{aligned} \tfrac{1}{2\gamma^2}\langle\Phi^2\rangle = &\left(2\textstyle\int_\tau^{2\tau} dt_2 \int_\tau^{t_2} dt_1 - \int_\tau^{2\tau} dt_2 \int_0^\tau dt_1\right) \\ &\cdot \langle\delta H[\mathbf{r}(t_2)]\delta H[\mathbf{r}(t_1)]\rangle. \end{aligned} \tag{7.47}$$

The average over diffusion paths is

$$\begin{aligned} &\langle\delta H[\mathbf{r}(t_2)]\delta H[\mathbf{r}(t_1)]\rangle \\ &\quad = \tfrac{1}{V_p}\textstyle\int_{V_p} dV_1 \int_{V_p} dV_2\, G(\mathbf{r}_2, \mathbf{r}_1|t_2 - t_1)\delta H(\mathbf{r}_2)\delta H(\mathbf{r}_1). \end{aligned} \tag{7.48}$$

There are two sources for the field deviation δH:

$$\delta H(\mathbf{r}) = \mathbf{r}\cdot\nabla H + \delta H_{\text{susc}}(\mathbf{r}), \tag{7.49}$$

where the first term represents a uniform, time independent gradient of the spin aligning magnetic field, while the second term is the result of the fluid-matrix magnetic susceptibility difference $\Delta\chi \equiv \chi_p - \chi_m$ (χ_p, χ_m are the magnetic susceptibilities of pore fluid and rock matrix, respectively). The latter is the primary source of field heterogeneities at the pore scale.

Susceptibility difference contribution

The portion of δH that is due to the magnetic susceptibility contrast is given by [19]

$$\delta H_{\text{susc}}(\mathbf{r}) \cong 4\pi\Delta\chi H_0 \sum_{\mathbf{g}\neq 0} \cos^2(\mathbf{g},\mathbf{e}_z)\theta_{\mathbf{g}} e^{i\mathbf{g}\cdot\mathbf{r}}. \tag{7.50}$$

Here $\theta_{\mathbf{g}} \equiv \frac{1}{V}\int dV\, \theta_p(\mathbf{r})e^{-i\mathbf{g}\cdot\mathbf{r}}$ [see Eq. (A.8)] is the $\mathbf{g}$-Fourier expansion coefficient of $\theta_p(\mathbf{r})$, and $\theta_p(\mathbf{r})$ is the characteristic function of the pore space, equal to 1 when $\mathbf{r}$ is inside that space and to 0 otherwise. The sum is over all the nonzero vectors $\mathbf{g}$ of the reciprocal lattice appropriate to the spatial periodicity of the pore space, and $(\mathbf{g},\mathbf{e}_z)$ represents the angle between $\mathbf{g}$ and the direction of the spin polarizing field, i.e., the z axis.

When this is substituted in the evaluation of $\langle\Phi^2\rangle$, we get the contribution due to the magnetic susceptibility contrast between pore fluid and solid matrix

$$\langle\Phi^2\rangle_{\text{susc}} = \frac{2}{\phi}(4\pi\gamma\Delta\chi H_0)^2 \sum_{n>0} b_n(\tau)|a_n|^2. \tag{7.51}$$

Here $b_n(\tau)$, a_n are quantities that depend on the $\mathbf{q}=0$ eigenstate of the band n: $b_0(\tau)\equiv 0$—that is why the $n=0$ term is absent in the sum—while $b_n(\tau)$ for $n\neq 0$ has a simple explicit form with useful limits for extreme values of τ [19]:

$$\begin{aligned} b_n(\tau) &= \frac{2\tau}{\lambda_{n0}} - \frac{1}{\lambda_{n0}^2}(1-e^{-\lambda_{n0}\tau})(3-e^{-\lambda_{n0}\tau}) \\ &\cong \begin{cases} \frac{2}{3}\lambda_{n0}\tau^3 & \text{when} \quad \lambda_{n0}\tau \ll 1, \\ \frac{2\tau}{\lambda_{n0}} & \text{when} \quad \lambda_{n0}\tau \gg 1. \end{cases} \end{aligned} \tag{7.52}$$

The a_n's depend on the $\mathbf{q}=0$ eigenfunctions $\phi_{n\mathbf{0}}(\mathbf{r})$ and are dimensionless:

$$a_n = \sum_{\mathbf{g}\neq 0} \cos^2(\mathbf{g},\mathbf{e}_z)\theta^*_{\mathbf{g}}\widetilde{\psi}_{n\mathbf{0}}(\mathbf{g}), \tag{7.53}$$

where $\widetilde{\psi}_{n\mathbf{0}}(\mathbf{g})$ is the $\mathbf{g}$-Fourier expansion coefficient of $\theta_p(\mathbf{r})\phi_{n\mathbf{0}}(\mathbf{r})$ [see Eq. (A.7)].

We note that, if the pore space is macroscopically isotropic, a_0 has a value that is independent of the detailed shape of that space [this is a consequence of the exact result $\phi_{0\mathbf{0}}(\mathbf{r}) = 1/\sqrt{\phi}$—see Eq. (A.13)]

$$a_0 = \frac{1}{\sqrt{\phi}} \sum_{\mathbf{g}\neq 0} \cos^2(\mathbf{g},\mathbf{e}_z)|\theta_{\mathbf{g}}|^2 = \frac{1}{3}\sqrt{\phi}(1-\phi).$$

The sum of all the other $|a_n|^2$ then satisfies the following inequality [19]:

$$\sum_{n>0} |a_n|^2 = \sum_{n} |a_n|^2 - a_0^2 \leq \frac{1}{3}\phi(1-\phi)\frac{2+\phi}{3}.$$

Note that the lowest eigenstate $n = 0$, $\mathbf{q} = 0$ does not contribute to the sum in Eq. (7.51).

Fixed gradient contribution

The contribution from an externally applied, fixed, uniform magnetic field gradient $\nabla H \parallel z$ is obtained by the following calculation

$$\begin{aligned}\tfrac{1}{\gamma^2}\langle\Phi^2\rangle_{\text{grad}} &= \left(2\textstyle\int_\tau^{2\tau} dt_2 \int_\tau^{t_2} dt_1 - \int_\tau^{2\tau} dt_2 \int_0^\tau dt_1\right) \\ &\quad \left(-\tfrac{1}{V_p}\right) \textstyle\int_{V_p} dV_1 \int_{V_p} dV_2 [(\mathbf{r}_2 - \mathbf{r}_1)\cdot\nabla H]^2 G(\mathbf{r}_2, \mathbf{r}_1 | t_2 - t_1). \end{aligned} \tag{7.54}$$

The spatial integrals are evaluated by taking derivatives of the Fourier transform of G

$$\begin{aligned} -\tfrac{1}{V_p}\textstyle\int_{V_p} dV_1 \int_{V_p} dV_2 [(\mathbf{r}_2 - \mathbf{r}_1)\cdot\nabla H]^2 G(\mathbf{r}_2, \mathbf{r}_1 | t_2 - t_1) \\ = \left(\nabla H \cdot \tfrac{\partial}{\partial \mathbf{k}}\right)^2 M(\mathbf{k}, t_2 - t_1)\Big|_{\mathbf{k}=0}, \end{aligned} \tag{7.55}$$

where

$$M(\mathbf{k}, t) \equiv \frac{1}{V_p}\int_{V_p} dV \int_{V_p} dV' G(\mathbf{r}, \mathbf{r}' | t) e^{-i\mathbf{k}\cdot(\mathbf{r}-\mathbf{r}')} \tag{7.56}$$

is the PFGSE amplitude.

The resulting additive contribution to the mean square of Φ from the fixed magnetic field gradient is

$$\langle\Phi^2\rangle_{\text{grad}} = (\gamma\nabla H)^2 \left(\frac{4}{3}D_e\tau^3 + \frac{1}{\phi}\sum_{n>0} b_n(\tau) d_n\right), \tag{7.57}$$

where d_n is given by [19]

$$d_n = \left(\gamma\nabla H \cdot \frac{\partial}{\partial \mathbf{q}}\right)^2 |\tilde{\psi}_{n\mathbf{q}}(\mathbf{0})|^2\Big|_{\mathbf{q}=0}.$$

From Eq. (7.52) it follows that, for very long times, only the first term of Eq. (7.57) is important and we get a result like Eq. (7.1), but with D_p replaced by

D_e. For intermediate times, the second term makes a significant contribution to the decay rate of the spin echo amplitude. That contribution depends on the $\mathbf{q} = 0$ diffusion eigenstates, and hence on the microstructure and on the value of D_p, in a non-trivial fashion. At very short times, the second term in Eq. (7.57) combines with the first term to give back the result of Eq. (7.1).

Summary of the theory

The spin echo decay rates, which result from putting the results of Eqs. (7.51) and (7.57) in the exponent of $e^{-\frac{1}{2}\langle\Phi^2\rangle}$, appear on top of the decay rate of the total spin polarization $M(t)$, which is present even if the polarizing field is completely uniform.

Eqs. (7.51) and (7.57), together with the second line of Eq. (7.52), show that at long times (i.e., $\lambda_{n0}\tau \gg 1$), a spectrum of exponential decays of the spin echo amplitude is produced, which depends on the diffusion eigenmodes. It therefore reflects details of the microstructure of the pore space. The precise relation between this distribution of decay rates and the microstructure is a complex one, even in the relatively simple case of periodic microstructures: Both the eigenvalues and the eigenfunctions of the diffusion problem participate in producing that relation.

Attempts were made, at times, to calculate the effects of susceptibility difference by deriving an average value for the spatially fluctuating field gradient, and using that in Eq. (7.1). That approach totally ignores the effects of the restricted geometry and the interface absorption on the diffusion process. In practice, this ad-hoc approximation appears to grossly overestimate the decay rate of the spin echos [80]. A somewhat better ad-hoc approximation is to consider the pore space as a collection of independent small pores and compute the field gradient in each one. A value of $\langle\Phi^2\rangle$ is then calculated for each pore separately, using Eq. (7.1). The results for individual pores are then averaged over the distribution of pore sizes and shapes—see Sections 4.2.3 and 4.2.4. This leads to much better agreement with experiments, though it relies heavily on a Gaussian assumption for the distribution of Φ^2 [80].

Clearly, the approach described here is better founded and more systematic. In particular, it allows the validity of the Gaussian approximation to be monitored, and it even allows one to go beyond that approximation. Its main disadvantage is the limitation to periodic microstructures.

Enhanced T_2 relaxation rate in CPMG experiments

The magnetic field inhomogeneities affecting T_2 measurements for porous media come from two sources. One is the field gradient externally applied to the system. It occurs over a much larger scale than the pore dimensions, and in the treatment described here we assume that it is uniform throughout the system. Another source of field inhomogeneities is the magnetic susceptibility difference between the pore fluid and solid matrix, which makes the *local field* non-uniform even when the *applied field* is accurately uniform. The resultant spatial variations are at the pore size scale and can be strongly affected by the local pore geometry. Assuming the phase angles of spins have a Gaussian distribution (which is a good approximation if $\langle\Phi^4\rangle_c \ll \langle\Phi^2\rangle$), the magnetization of the system at time 2τ is given by

$$\frac{M(2\tau)}{M(0)} = \exp\left(-\frac{2\tau}{T_2} - \frac{1}{2}\langle\Phi^2\rangle_{\rm susc} - \frac{1}{2}\langle\Phi^2\rangle_{\rm grad}\right), \tag{7.58}$$

where $\langle\Phi^2\rangle_{\rm susc}$ and $\langle\Phi^2\rangle_{\rm grad}$ are given by (7.51) and (7.57), respectively.

In a CPMG experiment, instead of just one π pulse, an entire sequence of π pulses is applied, with a time interval 2τ between sequential pulses. A corresponding train of echoes is observed, with gradually decreasing magnitudes [201]. For a periodic microstructure, it can be shown that the decay of that echo train is described by [19, 43]

$$\begin{aligned}\ln\frac{M(2n\tau)}{M(0)} + \frac{2n\tau}{T_2} &= -(\gamma\nabla H)^2\left[\frac{2n\tau^3 D_e}{3} + \frac{1}{2\phi}\sum_{m>0} c_m(\tau)\frac{\partial^2}{\partial q^2}\left|\tilde{\psi}_{m\mathbf{q}}(\mathbf{0})\right|^2\right]_{\mathbf{q}=0} \\ &\quad -\frac{1}{\phi}(4\pi\gamma\Delta\chi H_0)^2\sum_{m>0} c_m(\tau)|a_m|^2, \end{aligned}\tag{7.59}$$

where $c_m(\tau)$ is given by

$$c_m(\tau) = \frac{2n\tau}{\lambda_{m0}}\left(1 - \frac{\tanh(\lambda_{m0}\tau)}{\lambda_{m0}\tau}\right) + \frac{(1-e^{-\lambda_{m0}\tau})}{\lambda_{m0}^2(1+e^{-2\lambda_{m0}\tau})^2}[1+(-)^n e^{-2n\lambda_{m0}\tau}]. \tag{7.60}$$

The enhancement of the $1/T_2$ relaxation rate is obtained by differentiating Eq. (7.59) with respect to elapsed time t, where $t = 2n\tau$, keeping all the other factors of τ fixed.

Numerical examples

Detailed calculations of $\langle\Phi^2\rangle$ have only been made for model periodic microstructures with a simple unit cell—see Appendix A and Ref. [19]. The inverse problem, starting from the distribution of decay rates and trying to reconstruct the microstructure, has not been tackled even for those simple models. Moreover, details of the microstructure are often not important in themselves, but only inasmuch as they enable us to predict other macroscopic properties of the medium, such as permeability to fluid flow. That is why most attempts to connect diffusion properties and permeability have been empirical.

These calculations show that, under conditions that usually prevail in brine saturated porous rocks, $\frac{1}{24}\langle\Phi^4\rangle_c$ is less than $\frac{1}{2}\langle\Phi^2\rangle_c$. These statements are based upon results of calculations on periodic pore structures, like the ones shown in Figs. 7.3 and 7.4: Fig. 7.3 shows results of computations of $\frac{1}{2}\langle\Phi^2\rangle_c$ and $\frac{1}{24}\langle\Phi^4\rangle_c$, due to susceptibility difference $\Delta\chi$ between water and rock, for two representative values of that difference and for different values of a and τ. However, it is not uncommon to find brine-saturated rocks that contain strong paramagnetic impurities resulting in a large magnetic susceptibility contrast, and leading to non-Gaussian situation.

Fig. 7.4 shows similar plots due to a fixed field gradient ∇H. In both cases, the dependence of $\langle\Phi^2\rangle_c$ on $\Delta\chi$ or ∇H is *quadratic*, while the dependence of $\langle\Phi^4\rangle_c$ is *quartic*. That is why, when ∇H is increased from 1 Gauss/cm to 20 Gauss/cm, the relative importance of $\frac{1}{24}\langle\Phi^4\rangle_{c,grad}$ increases sufficiently so as to make it greater than $\frac{1}{2}\langle\Phi^2\rangle_{c,grad}$ for times greater than about 30 msec—see Fig. 7.4.

7.2 PFGSE and its applications

The pulsed-field-gradient-spin-echo (PFGSE) NMR method was first proposed in 1963 [157] and demonstrated experimentally in 1965 [205]. Basically, it is a spin echo experiment in which a uniform polarizing field H_0 is modified by applying the same uniform field gradient ∇H for a number of short time intervals during both the decay period and the reconstruction period of the NMR signal. Those field gradient pulses are set so that they would have a null effect on any precessing spin that does not change its position during the entire train of field gradient pulses. Thus, by comparing the spin echo

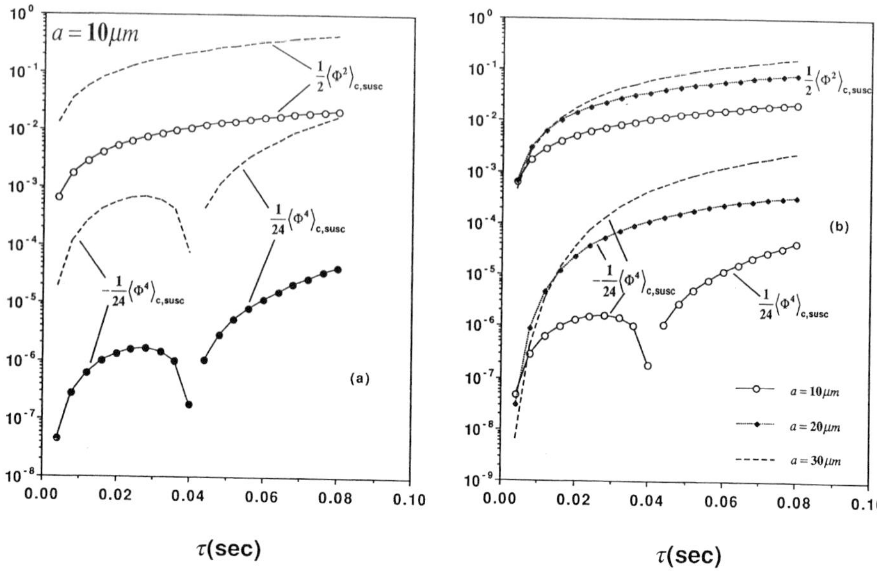

Figure 7.3: Results of $\frac{1}{2}\langle\Phi^2\rangle_{c,\text{susc}}$ and $\frac{1}{24}\langle\Phi^4\rangle_{c,\text{susc}}$ vs. τ for a brine filled periodic pore space made by embedding a simple cubic array of identical touching solid spheres of diameter a in the fluid. (a) $a = 10$ μm, $|\Delta\chi| = 0.6 \times 10^{-6}$emu/cm^3 (circles) and $|\Delta\chi| = 2.7 \times 10^{-6}$emu/cm^3 (dashed lines). (b) $|\Delta\chi| = 0.6 \times 10^{-6}$emu/cm^3 and three different values of a. Note that $\frac{1}{24}\langle\Phi^4\rangle_{c,\text{susc}}$ is positive for $a = 10$ μm when τ is greater than 40 ms, and negative for all other values of a and τ (after Fig. 4 of Ref. [19]).

with and without those field gradient pulses, information is obtained about motion of the precessing spins.

In the simplest type of PFGSE experiment, a single pulse of spatially uniform field gradient ∇H is first applied, for a short time interval δ between the initial rf $\pi/2$ pulse and the subsequent spin inverting rf π pulse. Then another identical ∇H pulse is applied between that π pulse and the subsequent spin echo. Fig. 2.9 shows a schematic representation of these events plotted vs. time (note that $\mathbf{g}$ is used to denote the field gradient ∇H in that figure).

The analysis of this experiment is based upon Eq. (7.42): The extra phase shift acquired by a single precessing spin is given by an integral over the deviation δH of the local polarizing field from its uniform background value $H_0 \parallel z$. That deviation now depends on time, because ∇H is turned

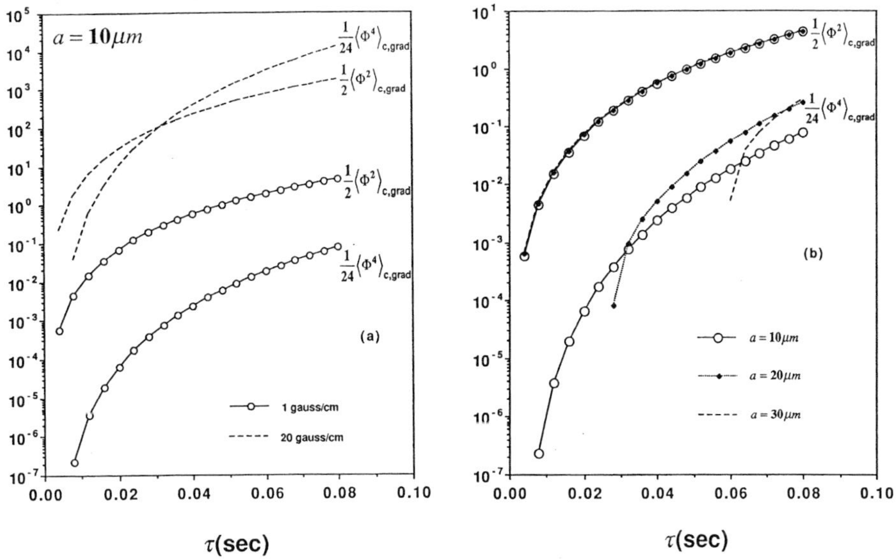

Figure 7.4: Results of $\frac{1}{2}\langle\Phi^2\rangle_{c,\text{grad}}$ and $\frac{1}{24}\langle\Phi^4\rangle_{c,\text{grad}}$ vs. τ for the same brine filled pore space as in Fig. 7.3, with an externally applied magnetic field gradient ∇H. (a) $a = 10$ μm and two different values of magnetic field gradient. (b) one magnetic field gradient (1 Gauss/cm) for different values of a (after Fig. 5 of Ref. [19]).

on only for two short time intervals, of length δ each, but also because the spin changes its position $\mathbf{r}(t)$ with time. If the change in position during the time δ is negligible, then that phase shift is given by

$$\begin{aligned}\Phi &= -\gamma\int_0^\tau dt\,\delta H[\mathbf{r}(t),t] + \gamma\int_\tau^{2\tau} dt\,\delta H[\mathbf{r}(t),t] \\ &\cong -\gamma\delta\left(\delta H[\mathbf{r}(t_1),t_1] - \delta H[\mathbf{r}(t_2),t_2]\right) = \gamma\delta\nabla H\cdot[\mathbf{r}(t_2)-\mathbf{r}(t_1)] \\ &\equiv \mathbf{k}\cdot[\mathbf{r}(t_2)-\mathbf{r}(t_1)], \qquad \mathbf{k} = \gamma\delta\nabla H,\end{aligned}$$

where t_1, t_2 are the moments at which the two gradient pulses are applied—in Fig. 2.9 the interval between those times is denoted by Δ. Note that the opposite signs of the phase shifts, caused by the two gradient pulses, is due to the fact that the quantum expectation value of $\hat{s}_x + i\hat{s}_y$, in the frame of reference which rotates around $H_0 \parallel z$ at the Larmor angular frequency γH_0, is changed into its complex conjugate by the spin inverting π pulse which is applied between the two gradient pulses—see Fig. 2.9 and the discussion in

Section 7.1.3. Obviously, if $\mathbf{r}(t_2) = \mathbf{r}(t_1)$ then the phase shifts caused by the two field gradient pulses cancel each other. Any non-perfect cancellation is due to spatial motion of the spin between the times t_1 and t_2.

The total effect on the spin echo signal from diffusive motion of all the precessing spins is found by averaging $e^{-i\Phi}$ over the distribution of diffusion paths $\mathbf{r}(t)$. This is done with the help of the diffusion propagator $G(\mathbf{r}, \mathbf{r}'|t)$, which is the probability density that a particle, which started out at $\mathbf{r}$, is found at $\mathbf{r}'$ after time t (that time is denoted by Δ in Fig. 2.9):

$$\langle e^{-i\Phi} \rangle = \frac{1}{V_p} \int_{V_p} dV \int_{V_p} dV' G(\mathbf{r}, \mathbf{r}'|t) e^{-i\mathbf{k}\cdot(\mathbf{r}-\mathbf{r}')} \bigg|_{t=|t_2-t_1|} \equiv M(\mathbf{k}, |t_2 - t_1|). \tag{7.61}$$

More complicated sequences of field gradient pulses have also been used, at times, in order to achieve a better signal-to-noise ratio in measurements, and also for achieving other specialized goals [114].

Eq. (7.61) is just the time dependent spatial Fourier transform of the average diffusion propagator. $M(\mathbf{k}, t)$ is also known as the "PFGSE amplitude", and can obviously be measured for values of $\mathbf{k}$ and t that are chosen by setting the experimental parameters. This capability has enormous potential for providing detailed information about the microstructure of the pore space, with a resolution that is determined by the range of $\mathbf{k}$ values that are accessed.

7.2.1 Diffraction-like behavior of PFGSE

Callaghan et al. [50] demonstrated very convincingly the diffraction-like behavior of PFGSE in an NMR experiment on a water filled porous medium, made by randomly and loosely packing a collection of monodispersed (i.e., same size) polystyrene spheres of diameter 16 μm. In this experiment, a peak in $M(\mathbf{k}, t)$ was found to develop, as the time $t = t_2 - t_1$ increased up to 110 msec, at a value of $|\mathbf{k}|$ corresponding to a pore separation distance $2\pi/|\mathbf{k}|$ of about 16 μm. In fact, this diffraction-like behavior can be shown theoretically to be a unique property of $M(\mathbf{k}, t)$. We shall briefly discuss some methods for calculating the PFGSE amplitude and some results of such calculations.

In Appendix A we show that, for a periodic pore space, the differential equation for the $\rho = 0$ diffusion eigenstates can be recast as a discrete matrix eigenvalue problem, where the eigenvectors are the Fourier expansion coefficients of the diffusion eigenfunctions—see Eq. (A.3). Since there are infinitely many expansion coefficients, the matrix is discrete but infinite. By restricting the calculation to a finite subset of those expansion coefficients, the infinite matrix gets truncated to a finite size matrix, whose eigenvectors and eigenvalues can be found by standard numerical procedures. The eigenstates of the $\rho \neq 0$ diffusion problem are then expanded in the $\rho = 0$ eigenstates, and this leads to a discrete matrix eigenvalue problem for the $\rho \neq 0$ eigenstates. A truncated version of the latter problem is then solved using standard numerical procedures.

To compute the PFGSE amplitude, $M(\mathbf{k}, t)$, we recall that the diffusion propagator has the following general form—see Eq. (7.26):

$$G(\mathbf{r}, \mathbf{r}'|t) = \frac{1}{V} \sum_{\lambda} e^{-\lambda t} \psi_\lambda(\mathbf{r}) \psi_\lambda^*(\mathbf{r}').$$

When this is substituted in Eq. (7.61), we get

$$M(\mathbf{k}, t) = \frac{1}{\phi} \sum_{n} e^{-\lambda t} |\widetilde{\psi}_\lambda(\mathbf{k})|^2, \tag{7.62}$$

where

$$\widetilde{\psi}_\lambda(\mathbf{k}) = \frac{1}{V} \int_V dV\, \theta_p(\mathbf{r}) \psi_\lambda(\mathbf{r}) e^{-i\mathbf{k}\cdot\mathbf{r}} = \frac{1}{V} \int_{V_p} dV\, \psi_\lambda(\mathbf{r}) e^{-i\mathbf{k}\cdot\mathbf{r}}$$

is the Fourier transform of $\theta_p(\mathbf{r})\psi_\lambda(\mathbf{r})$, with $\psi_\lambda(\mathbf{r})$ normalized as in Eq. (7.29).

Periodic media

For periodic porous media, the diffusion propagator becomes [compare this with Eq. (7.46), which refers to the special case where $\rho = 0$]

$$G(\mathbf{r}, \mathbf{r}'|t) = \frac{1}{V} \sum_{n\mathbf{q}} e^{-\mu_{n\mathbf{q}}|t|} \psi_{n\mathbf{q}}(\mathbf{r}) \psi_{n\mathbf{q}}^*(\mathbf{r}') e^{i\mathbf{q}\cdot(\mathbf{r}-\mathbf{r}')},$$

where $\mu_{n\mathbf{q}}$ is an eigenvalue and $\psi_{n\mathbf{q}}(\mathbf{r})$ is the periodic part of the $\rho \neq 0$ Bloch-Floquet-type eigenfunction, normalized as in Eq. (7.29)—see Appendix A. $M(\mathbf{k}, t)$ is then given by the following expression

$$M(\mathbf{k}, t) = \frac{1}{\phi} \sum_{n} e^{-\mu_{n\mathbf{q}} t} |\widetilde{\psi}_{n\mathbf{q}}(\mathbf{g}_\mathbf{k})|^2, \quad \mathbf{q} = \mathbf{k} - \mathbf{g}_\mathbf{k}, \tag{7.63}$$

where $\mathbf{g_k} \equiv \mathbf{k} - \mathbf{q}$ is the reciprocal lattice vector that is closest to $\mathbf{k}$, and

$$\widetilde{\psi}_{n\mathbf{q}}(\mathbf{g}) \equiv \frac{1}{V_a} \int_{V_p \cap V_a} dV \, \psi_{n\mathbf{q}}(\mathbf{r}) e^{-i\mathbf{g}\cdot\mathbf{r}}, \qquad \sum_{\mathbf{g}} \widetilde{\psi}^*_{n\mathbf{q}}(\mathbf{g}) \widetilde{\psi}_{m\mathbf{q}}(\mathbf{g}) = \delta_{nm} \tag{7.64}$$

are the normalized eigenvectors of the diffusion problem. Also, V_a is the volume of a single unit cell of the periodic porous medium, including both pore space and rock matrix, while the integral is only over the volume of pore space in one unit cell. This expression for $M(\mathbf{k}, t)$ is valid irrespective of whether $\rho = 0$ or $\rho \neq 0$, even though the values of $\mu_{n\mathbf{q}}$ and $\widetilde{\psi}_{n\mathbf{q}}(\mathbf{g})$, as well as the calculational procedure for obtaining these quantities, are quite different, as explained above. In the following, we will use the notation $\lambda_{n\mathbf{q}}$, $\phi_{n\mathbf{q}}(\mathbf{r})$ to denote the $\rho = 0$ limit of $\mu_{n\mathbf{q}}$, $\psi_{n\mathbf{q}}(\mathbf{r})$.

The extent to which $M(\mathbf{k}, t)$ includes detailed information about the microstructure is exhibited dramatically in the case of a periodic microstructure. Even though $G(\mathbf{r}, \mathbf{r}'|t)$ is not a periodic function, its Fourier transform $M(\mathbf{k}, t)$ exhibits peaks and valleys at characteristic values of $\mathbf{k}$ that are intimately connected to the spatial periodicity. Fig. 7.5 shows this behavior for $\mathbf{k}$ along the (100) direction in a microstructure which results when a simple cubic array of touching solid spheres is embedded in an otherwise uniform fluid, assuming $\rho = 0$. Fig. 7.6 shows similar results for the same system when $\rho a/D = 1$ and 10. The oscillatory behavior of $M(\mathbf{k}, t)$ is reminiscent of the periodic dependence of the eigenvalues $\mu_{n\mathbf{q}}$ upon $\mathbf{q}$ [see Figs. A.1 and A.2 in Appendix A], which is due to the periodic microstructure and follows from Bloch's theorem for such structures.

Properties of $M(\mathbf{k}, t)$

There are several things worth noting in connection with $M(\mathbf{k}, t)$ and Fig. 7.5: (a) The PFGSE amplitude is real, and it is even under space inversion, namely $M(-\mathbf{k}, t) = M(\mathbf{k}, t)$. This follows from the fact that G is real and symmetric $G(\mathbf{r}, \mathbf{r}'|t) = G(\mathbf{r}', \mathbf{r}|t)$. (b) From the inversion symmetry of the microgeometry it follows that

$$\theta_{\mathbf{g}} = \theta_{-\mathbf{g}} = \text{real} \tag{7.65}$$

$$\lambda_{n-\mathbf{q}} = \lambda_{n\mathbf{q}} \tag{7.66}$$

$$\widetilde{\psi}_{n-\mathbf{q}}(\mathbf{g}) = \pm\widetilde{\psi}_{n\mathbf{q}}(-\mathbf{g}) = \text{real.} \tag{7.67}$$

As a result, we only need to compute eigensolutions for $|\mathbf{q}|a$ from 0 up to the first Brillouin zone boundary π for any direction of $\mathbf{q}$, and retain all the

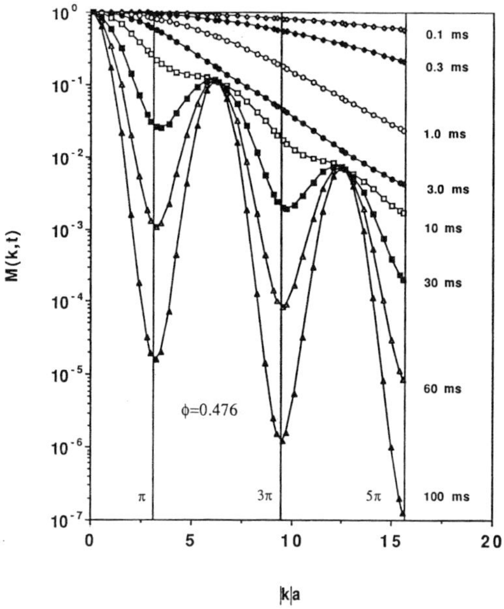

Figure 7.5: The PFGSE amplitude $M(\mathbf{k},t)$ as function of $|\mathbf{k}|a$ along the (100) direction at various times for a simple cubic array of identical touching spheres with ϕ=0.476 and $\rho = 0$. The Bragg planes at π, 3π, and 5π are also indicated (after Fig. 7 of Ref. [77]).

necessary weights $|\tilde{\psi}_{n\mathbf{q}}(\mathbf{g_k})|^2$ in addition to $|\tilde{\psi}_{n\mathbf{q}}(\mathbf{0})|^2$. This allows us to plot $M(\mathbf{k},t)$ beyond the first Brillouin zone boundary, in fact for any $\mathbf{k} = \mathbf{g_k} + \mathbf{q}$. (c) For sufficiently long times t, $M(\mathbf{k},t)$ develops peaks as function of $\mathbf{k}$ whenever $\mathbf{q} \equiv \mathbf{k} - \mathbf{g_k} = 0$. The values of $M(\mathbf{k},t)$ at those peaks, once they have developed, are independent of time. This is because there is a zero eigenvalue $\lambda_{0\mathbf{0}} = 0$ with a nonzero weight $|\tilde{\psi}_{0\mathbf{0}}(\mathbf{g_k})|^2 \neq 0$, which becomes the only important term in Eq. (7.63) for sufficiently long times. At other values of $\mathbf{k}$, all the eigenvalues with nonzero weights are greater than zero. Therefore, $M(\mathbf{k},t)$ continues to decay with increasing t. (d) When t is large enough so that the above mentioned peaks develop, it is also large enough that the lowest eigenvalue $\lambda_{0\mathbf{q}}$ becomes the dominant term in Eq. (7.63) for any other $\mathbf{k}$. In that case, a minimum of $M(\mathbf{k},t)$ usually appears whenever $\mathbf{q} \equiv \mathbf{k} - \mathbf{g_k}$ approaches the edge of the first Brillouin zone, because $\lambda_{0\mathbf{q}}$ has a maximum there. The position of this minimum is close to but not exactly at the zone edge, because the weight $|\tilde{\psi}_{0\mathbf{q}}(\mathbf{g_k})|^2$ also depends upon $\mathbf{q}$. (e) We can see from Fig. 7.5 that the lowest eigenvalue is the only one that needs

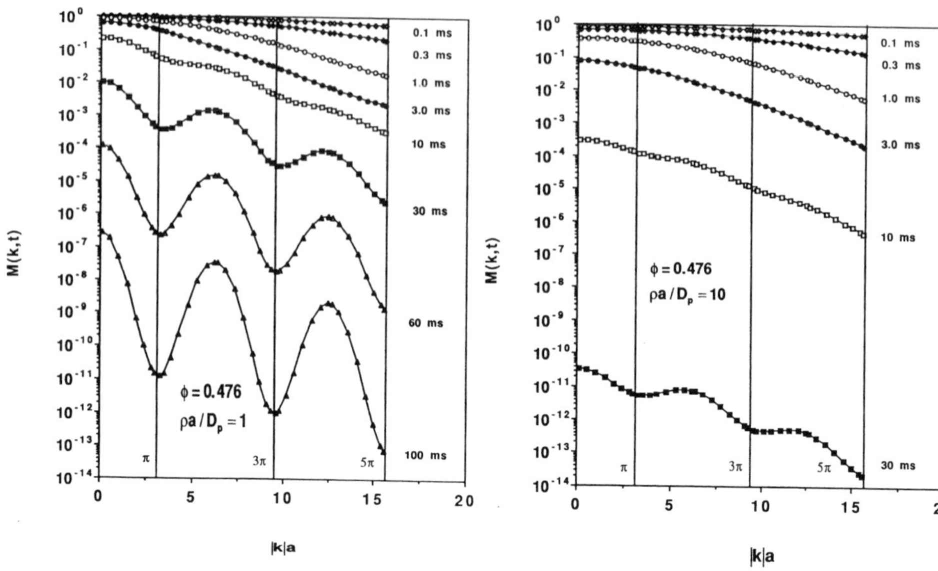

Figure 7.6: The PFGSE amplitude $M(\mathbf{k}, t)$ plotted vs. $|\mathbf{k}|a$ along the (100) direction at various times for a cubic array of touching spheres ($\phi = 0.476$) (a) for $\rho a/D_p = 1$ (b) for $\rho a/D_p = 10$ (after Fig. 6 of Ref. [18]).

to be considered as soon as $t \geq 0.01$ sec. That is 1/4 of the natural time scale of this diffusion problem $a^2/D_p = 0.04$ sec, obtained using $a = 10\,\mu$m for the unit cell size and $D_p = 2.5 \times 10^{-5}$ cm^2/sec. For such times the values of $M(\mathbf{k}, t)$ at the minima quickly become much smaller than the detection capability of any practical measuring device.

Fig. 7.6 exhibits qualitatively similar behavior for systems with nonzero ρ, but the oscillations of $M(\mathbf{k}, t)$ are increasingly damped with increasing ρ. The overall decay of $M(\mathbf{k}, t)$ with increasing time also increases with ρ.

Another feature which sometimes appears can be seen in Figs. 7.7 and 7.8, which show results for the same systems as in Figs. 7.5 and 7.6(a), but now with $\mathbf{k} \parallel (110)$. Near the middle of the second Brillouin zone, a sharp minimum appears near the position where a maximum is trying to develop at $\mathbf{k} = \mathbf{g}_{\mathbf{k}}$ with increasing t. The sharp minimum is caused by the fact that the weight factor $|\widetilde{\psi}_{0\mathbf{q}}(\mathbf{g}_{\mathbf{k}})|^2$ of the lowest eigenvalue band vanishes for some small value of $\mathbf{q} = \mathbf{k} - \mathbf{g}_{\mathbf{k}}$, i.e., for $\mathbf{k}$ very near to $\mathbf{g}_{\mathbf{k}}$. This feature persists even when $\rho \neq 0$. In the case where $\rho = 0$, we have $\widetilde{\psi}_{00}(\mathbf{g}) \propto \theta_{\mathbf{g}}$, therefore this feature is connected to the fact that $\theta_{\mathbf{g}_{\mathbf{k}}}$ is very small. Clearly,

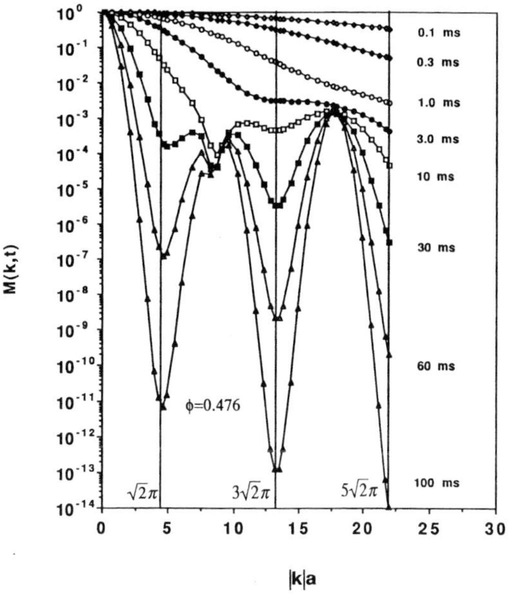

Figure 7.7: The PFGSE amplitude $M(\mathbf{k}, t)$ as function of $|\mathbf{k}|a$ along the (110) direction at various times for a simple cubic array of identical touching spheres with $\phi = 0.476$ and $\rho = 0$ (after Fig. 11 of Ref. [77]).

such an occurrence depends on the detailed shape of the pore space. This demonstrates the sensitivity of $M(\mathbf{k}, t)$ to those details, and indicates its potential usefulness for obtaining information about them.

7.2.2 Return to the origin probability (RTOP)

In Sections 7.1.2, 7.1.3 we saw that measurements of the time dependent diffusion coefficient $D(t)$ can provide information about the microstructure of the porous medium. That information came in the form of the value of S_p/V_p (from the short time behavior), the value of D_e (from the long time behavior), and the spectrum of relaxation rates (from the intermediate time behavior).

Another measurable quantity, with a similar potential for providing microstructural information about the porous medium, is the integrated return-to-the-origin probability (RTOP), defined as

$$P(t) \equiv \int_{V_p} dV G(\mathbf{r}, \mathbf{r}|t). \tag{7.68}$$

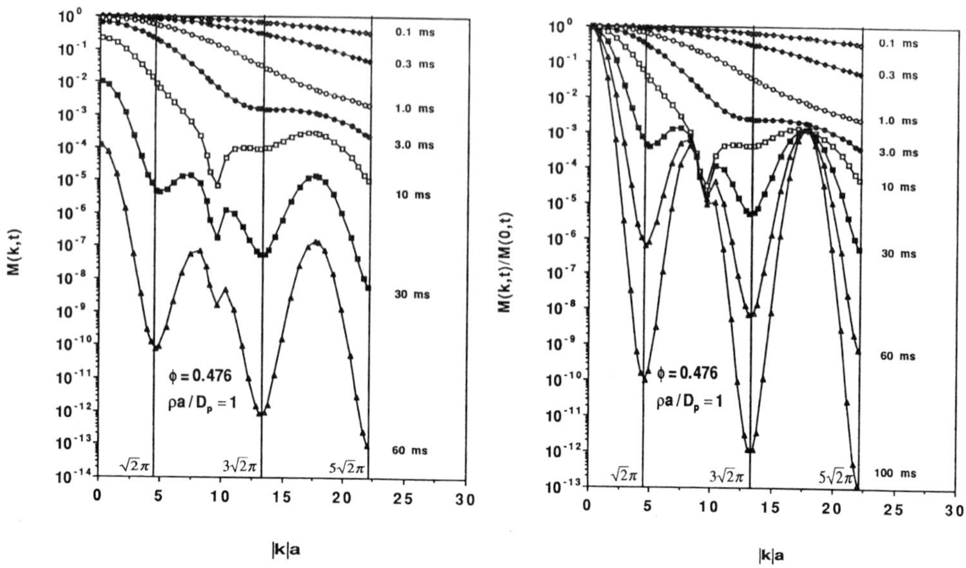

Figure 7.8: Plots of (a) $M(\mathbf{k}, t)$ and (b) $M(\mathbf{k}, t)/M(\mathbf{0}, t)$ vs. $|\mathbf{k}|a$ along the (110) direction at various times for a cubic array of touching spheres ($\phi = 0.476$) and $\rho a/D_p = 1$ (after Fig. 8 of Ref. [18]).

This definition is based upon the fact that $G(\mathbf{r}, \mathbf{r}|t)dV$ is the conditional probability of a particle to be found in the infinitesimal volume element dV near the point $\mathbf{r}$, at the time t, if it was at that same point $\mathbf{r}$ at time $t = 0$. When this is integrated over the entire pore space V_p, it becomes the *average* probability of any particle returning to its starting position after a time t. We shall see that the RTOP depends on the microstructure in ways that enable us to extract information about the microgeometry of the porous medium.

A more general concept is the average probability density for any particle to advance by the vector displacement $\mathbf{R}$ after a time t. That is the following volume average of the diffusion propagator (this is the same as the ensemble average in the case of a disordered microstructure), and is given by

$$P(\mathbf{R}, t) \equiv \frac{1}{V_p} \int dV G(\mathbf{R} + \mathbf{r}, \mathbf{r}|t) = \int \frac{d^3k}{(2\pi)^3} M(\mathbf{k}, t) e^{i\mathbf{k}\cdot\mathbf{R}}.$$

It is a function of t and of the "relative coordinate" $\mathbf{R}$. $P(\mathbf{R}, t)$ always starts out with the simple, well defined, delta function form

$$P(\mathbf{R}, t = 0) = \delta^3(\mathbf{R}),$$

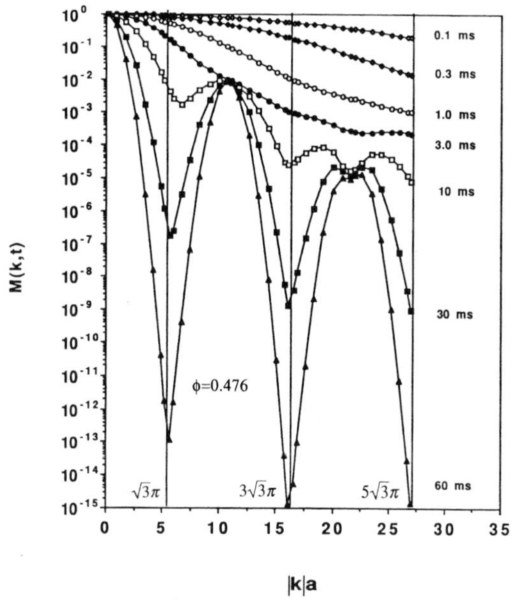

Figure 7.9: The PFGSE amplitude $M(\mathbf{k}, t)$ as function of $|\mathbf{k}|a$ along the (111) direction at various times for a simple cubic array of identical touching spheres with ϕ=0.476 and $\rho = 0$ (after Fig. 12 of Ref. [77]).

and ends up as $P(\mathbf{R}, t \to \infty) = 0$ for a connected pore space, or when there are nonzero spin absorption coefficients.

When $\mathbf{R} = \mathbf{0}$, this is the same as the integrated return to the origin probability (RTOP) $P(t)$, *divided by* V_p. We conclude that $P(t = 0)$ is infinite. However, for all $t > 0$, $P(t)$ is a finite, well defined quantity, which can be obtained from $M(\mathbf{k}, t)$:

$$P(t) = V_p P(\mathbf{R} = 0, t) = V_p \int \frac{d^3k}{(2\pi)^3} M(\mathbf{k}, t) = \sum_{\lambda} e^{-\lambda t}. \tag{7.69}$$

The last result follows from Eqs. (7.26) and (7.29), and shows that the RTOP is a "spectral function" of the diffusion problem in the porous medium: It is equal to a simple sum of all the characteristic exponential decay factors of the eigenstates of this problem.

If one could measure the RTOP with sufficient accuracy, one would be able to extract all the characteristic decay rates λ. It is amazing that such a simple arithmetic sum of exponentials also has the vivid physical interpretation of

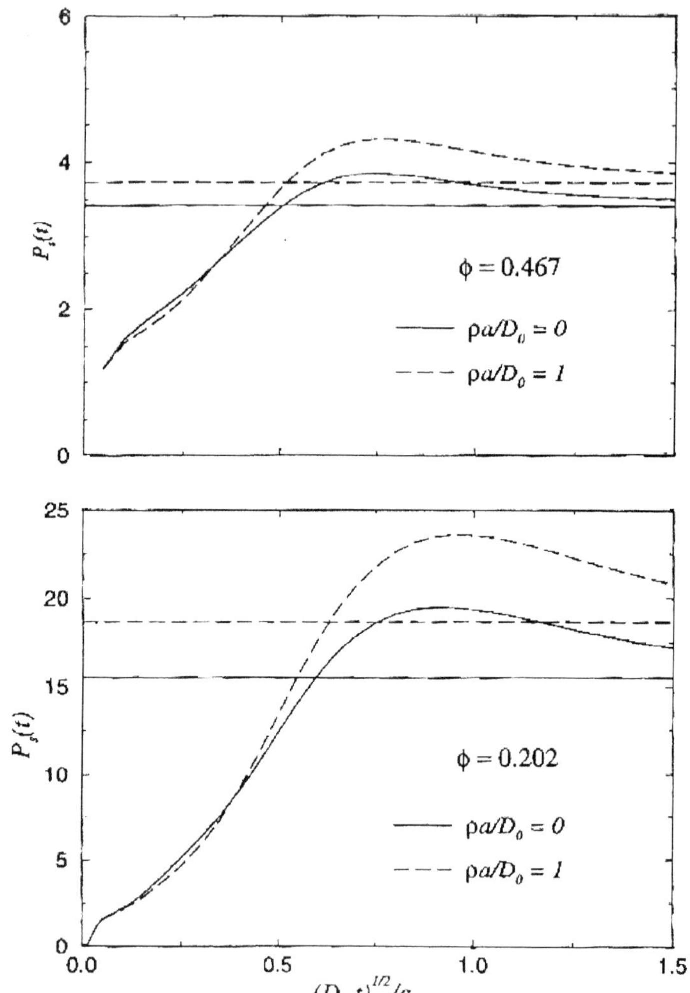

Figure 7.10: The effects of surface relaxation on the RTOP are illustrated for the touching sphere (upper panel) and overlapping sphere (lower panel) arrays of simple cubic packings. Note that, to leading order, the short time behavior of $P_s(t)$ is *not* affected by ρ. The horizontal straight lines represent the long time asymptotic result of Eq. (7.72) (after Fig. 8 of Ref. [192]).

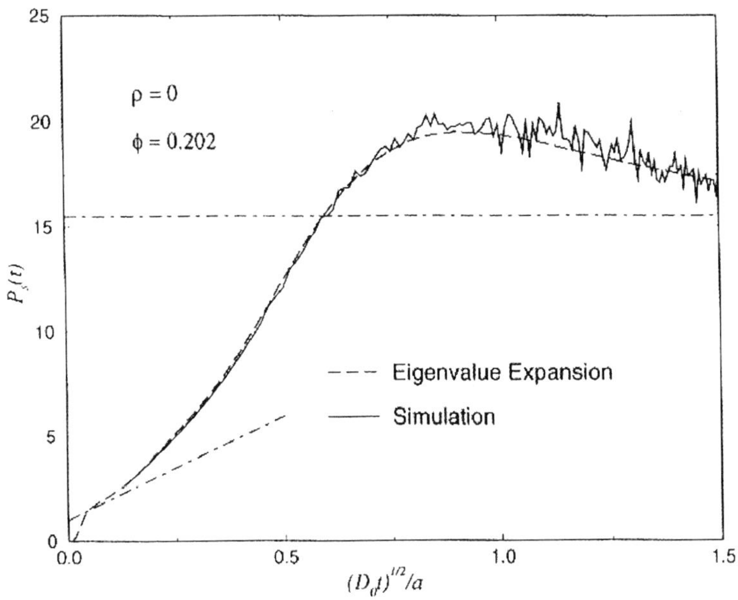

Figure 7.11: For a consolidated simple cubic packing of solid spheres (i.e., neighboring solid spheres overlap partially) with $\rho = 0$, eigenvalue expansion results for $P_s(t)$ are compared to partially smoothed data based on singular-value-decomposition analysis of the corresponding random walk simulations. Note that the random walk results fluctuate even after that drastic smoothing procedure. The lower (slanted) dot-dash line represents the short time asymptotic result of Eq. (7.71), while the upper (horizontal) dot-dash line represents the long time asymptotic result of Eq. (7.72) (after Fig. 4 of Ref. [192]).

At long times, both $D(t)$ and $P_s(t)$ tend to constant asymptotic values, which depend only on the lowest lying eigenstates [see Eqs. (7.35) and (7.72)]:

$$D(t) \to D_e = D_p \frac{\int_{V_p} dV \psi_0^2 (\nabla \phi_\alpha)^2}{\int_{V_p} dV \psi_0^2}, \quad P_s(t) \to \frac{1}{\phi} \left(\frac{D_p}{D_e}\right)^{3/2} \frac{V_p \int_{V_p} dV \psi_0^2(\mathbf{r})}{\left[\int_{V_p} dV \psi_0(\mathbf{r})\right]^2}.$$

All of the preceding considerations lead to the conclusion that the asymptotic behavior of a diffusing particle in a porous medium, at times that are either very short or very long, can provide only limited information about the microstructure. In order to obtain detailed information, the behavior of both $D(t)$ and $P_s(t)$ at intermediate times (which depend on the entire spectrum of eigenstates) must be measured, and their dependence on the microstructure topology must be studied and understood.

7.3 Random walker simulations

Diffusive motion of a particle is the result of many interactions which take place, between that particle and other neighboring particles, in an *incoherent fashion*, i.e., with rapid loss of memory as time progresses. This is often described by the simple model of "uncorrelated discrete random walk", in which time advances by a finite step τ (τ can either be fixed or variable from step to step), and the particle position at time $t + \tau$ is chosen at random from a set of possible positions according to a probability distribution.

Various implementations of this model have been used in computer simulations of a physical diffusion process. Such simulations mimic the actual microscopic processes in a simplified manner. Nevertheless, they allow us to follow the microscopic motion of the particle with some finite small scale of resolution.

Random walker simulations of diffusive motion have the advantage that, in contrast with solutions of the differential equation for the spin density, they are able to deal with a complex and disordered pore space almost as easily as when that space has a simple ordered shape.

The main disadvantage is that the simulation can proceed only for times that are not excessively long compared to the basic time step τ. In simulations where D_e is much less than D_p, either as a result of bottlenecks in the pore space or as a result of strong absorption at the pore/matrix interface, the limitation to short times has particularly disastrous effects on the attainable results: When narrow necks or cracks play a dominant role, it takes a very long time for the random walker to explore those elements of the pore space, and for their full impact to affect macroscopic quantities like $M(\mathbf{k}, t)$, $D(t)$, and $P(t)$. When ρ is large, many walkers get wiped out before they can explore the detailed shape of the pore space. Thus, many more walkers must be simulated in order to determine those macroscopic quantities.

Nevertheless, random walker simulations can be carried out to times that are well beyond those where short time asymptotics are valid—see Appendix B. They therefore allow quantitative comparisons to be made, not only with those asymptotics, but also with results based on numerical calculations of the diffusion eigenstates, which are valid only for times that are not too short—see Appendix A. The latter correspondence is especially important: On one hand, it allows the parameters of the random walk simulations to be chosen so as to correctly reproduce the results of the diffusion equation. On the other hand, by comparing random walk simulations on periodic and non-periodic pore structures, which have common elements like the typical pore sizes, narrow necks, and narrow cracks, we can determine the extent to which periodic models of the pore structure are able to mimic the macroscopic behavior of the disordered pore space found in real rocks.

Infinite fluid

The simplest computational implementation of a three dimensional random walk in an unrestricted geometry has a fixed time step τ, and a fixed simple cubic lattice of discrete possible positions in space, with lattice constant ϵ. Starting at some given position, the walker advances to one of the nearest neighbor positions, chosen at random and with equal probabilities, as the time advances by τ. After N time steps, the walker has reached a position some distance away from the origin. A simple statistical analysis shows that the *root mean square distance* R is given, for large N, by $R = \sqrt{N}\epsilon$. Since the total time elapsed is $t = N\tau$, we can determine D_p by the following expression, which is based upon Eq. (7.24):

$$D_p = \frac{R^2}{6t} = \frac{\epsilon^2}{6\tau}. \tag{7.73}$$

Note that the number of discrete time steps N has dropped out of the final result. It is worth noting that, in this simple model, all the above results are valid for *arbitrary values of* N, i.e., even if the number of steps N is as small as 1.

Restricted geometry

In a restricted geometry or pore space, the walker sometimes encounters a boundary, where it is either reflected or wiped out (mimicking interface absorption of the particle or depolarization of the spin). This can be implemented by modifying the above described algorithm as follows: When the walker is at a boundary site and tries to move into the boundary (i.e., in a direction where the nearest neighbor site is absent), it is either wiped out with probability γ, or it stays put at the boundary site with probability $1-\gamma$, even though the time advances by τ.

The various quantities $M(t)/M(0)$, $M(\mathbf{k},t)$, $D(t)$ can now be calculated from the discrete paths $\mathbf{r}_n(t)$, $n = 1 \ldots N$ followed by a sufficiently large number N of simulated random walkers: $M(t)/M(0)$ is simply equal to the fraction of walkers that survive at time t. The PFGSE amplitude is found from

$$M(\mathbf{k},t) = \frac{1}{N}\sum_{n=1}^{N} e^{i\mathbf{k}\cdot[\mathbf{r}_n(t)-\mathbf{r}_n(0)]}.$$

The time dependent diffusion coefficient can be found from this, using Eq. (7.37) or Eq. (A.16). However, it is easier to use the more physical definition of that coefficient, expressed by Eq. (A.14), to get

$$6D(t)t = \frac{1}{N}\sum_{n=1}^{N}[\mathbf{r}_n(t) - \mathbf{r}_n(0)]^2. \tag{7.74}$$

Note, however, that those two definitions of $D(t)$ lead to different results when $\rho \neq 0$—see the discussion which follows Eq. (A.15). Finally, $P(t)$ is found by integrating $M(\mathbf{k},t)$ over all $\mathbf{k}$, as described in Eq. (7.69) [192].

The relation between the continuum diffusion coefficient of the pore fluid D_p and the random walk parameters ϵ, τ, i.e., Eq. (7.73), has a very wide range of validity: It extends to other implementations of a random walk, e.g., to implementations where the walker position after time τ is not limited to a discrete set of nearest neighbors but can lie anywhere on a spherical surface of radius ϵ around the origin. It also extends to an implementation where the spatial step size ϵ and the time step τ are changed from step to step: At each step, ϵ is made as large as possible without getting any intersections of that spherical surface with any boundary or interface. At the same time, τ is also changed so as to keep satisfying Eq. (7.73) with a fixed value of D_p. It turns out that, as long as these restrictions are maintained, the random walk mimics the diffusion process faithfully [193]. The latter implementation can greatly accelerate the computation by allowing large steps to be used, whenever possible, in the random walker simulation.

Mimicking interface absorption

In contrast with the simplicity and widespread validity of Eq. (7.73), the relation between the continuum interface absorption coefficient ρ and the probability γ, that the walker is wiped out when it reaches the boundary of

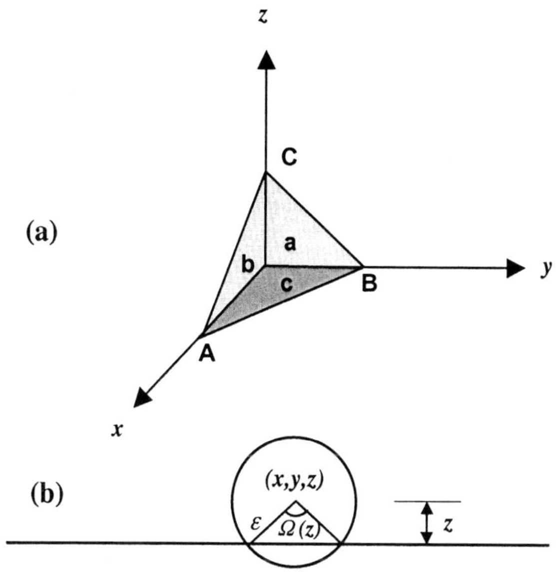

Figure 7.12: (a) Illustration of the surface area enhancement which results from a discretization of space in a random walker simulation. (b) Illustration of the configuration used to calculate the probability for the random walker to be wiped out using a random walk implementation where the walker jumps to an arbitrary point on the surface of a sphere of radius ϵ, as described briefly in the text, and in greater detail in Section IIA of Ref. [20] (after Fig. 6 of Ref. [20]).

the pore space, involves more subtle reasoning. This is most readily appreciated when the sites attainable by that walker are restricted to lie on a fixed simple cubic lattice. This spatial discretization has the effect of changing most smooth interfaces into a collection of right-angled corners. This is evident if we consider a situation where the interface intersects three neighboring orthogonal edges or bonds of the lattice—see Fig. 7.12(a). The discretization thus has the effect of increasing the area of the smooth interface by a factor that *depends on the local orientation of the interface* and is usually *significantly greater than* 1.

Other implementations of the random walk also enhance the interface area, by an amount whose average is usually non-negligible and depends on details of the implementation—see, e.g., Fig. 7.12(b). The consequent enhancement factors of the total interface area *do not tend to* 1 *as the step size of discretization is decreased*, but to a value significantly greater than 1.

For an interface which is not only macroscopically isotropic, but is made up entirely of closed spherical surfaces, the value of this enhancement factor in the discrete space implementation, averaged over all possible orientations of the interface, is 3/2. Thus the relation between γ and ρ is found to be [20]

$$\gamma = \frac{2}{3}\frac{\epsilon\rho}{D_p}.$$

When these subtleties are taken into consideration, random walker simulations provide an important tool for studying diffusion in a porous medium. In particular, they are able to study disordered pore structures and, for periodic microstructures, they are able to bridge any gap that remains between short time asymptotics and eigenstate based calculations, whose validity breaks down at short times. Figs. 7.13 and 7.14 show results for $M(\mathbf{k}, t)$, calculated for periodic simple cubic arrays of spherical obstacles of radius a using both methods, with different values of ϕ and with $\rho a/D_p = 0$, 0.2, and 1. A similar comparison for $D(t)$ appears in Fig. 7.15, and a similar comparison for the normalized return-to-the-origin-probability-density (RTOP) $P_s(t)$ appears in Fig. 7.11. These comparisons lend confidence in both computational methods, whose ranges of validity are largely complementary.

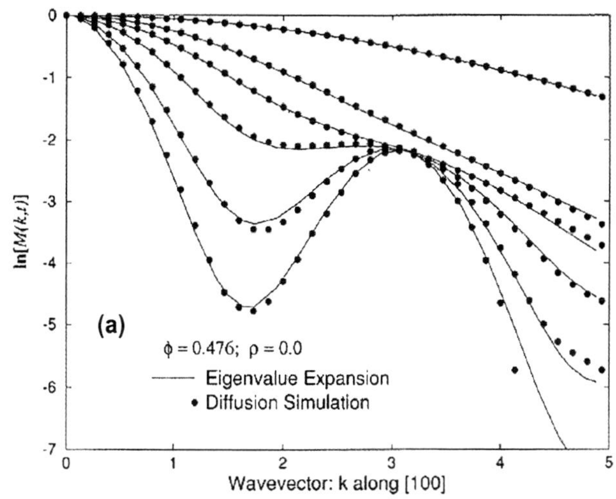

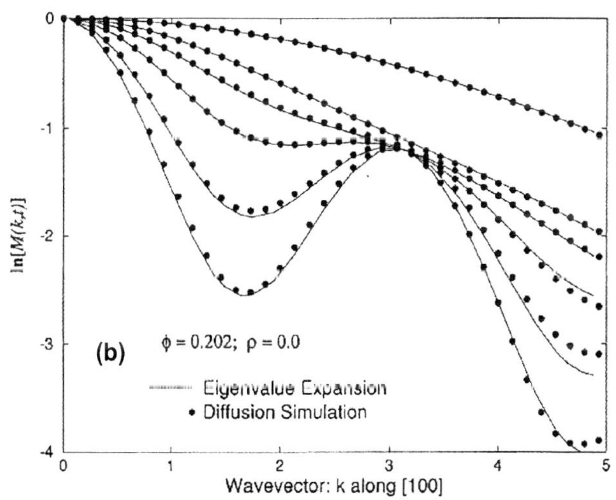

Figure 7.13: Eigenstate expansion and random walker simulation results for $M(\mathbf{k}, t)$, in a cubic array of spheres microstructure, are compared in the case $\rho = 0$ (a) for touching spheres ($\phi = 0.4674$) (b) for overlapping spheres ($\phi = 0.202$). Results are presented for six values of t corresponding to 0.67, 3.33, 6.67, 13.33, 26.67 and 40 msec, in descending order of the curves. The cubic unit cell edge is taken to be 10 μm and $D_p = 2.5\,\mu\text{m}^2/\text{msec}$ (after Fig. 2 of Ref. [20]).

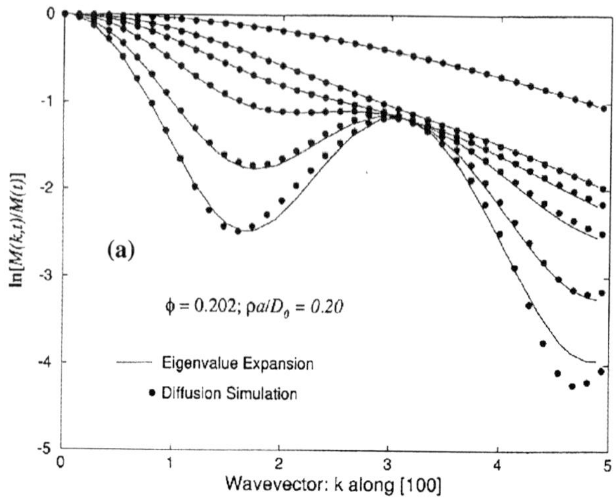

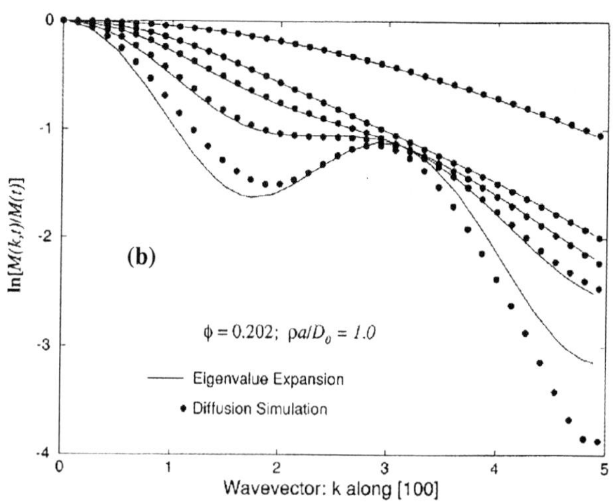

Figure 7.14: Comparison of results for $M(\mathbf{k},t)/M(t)$, using the eigenstate expansion and random walker simulations, in the same microstructure as that of Fig. 7.13. In this comparison, two different nonzero values were assumed for the relaxation strength ρ: In (a) $\rho a/D_p = 0.20$ and $t = 0.67 \to 40$ msec, while in (b) $\rho a/D_p = 1.0$ and $t = 0.67 \to 26.67$ msec. In both sets of calculations, $\phi = 0.202$; in the figure D_p is denoted by D_0 (after Fig. 4 of Ref. [20]).

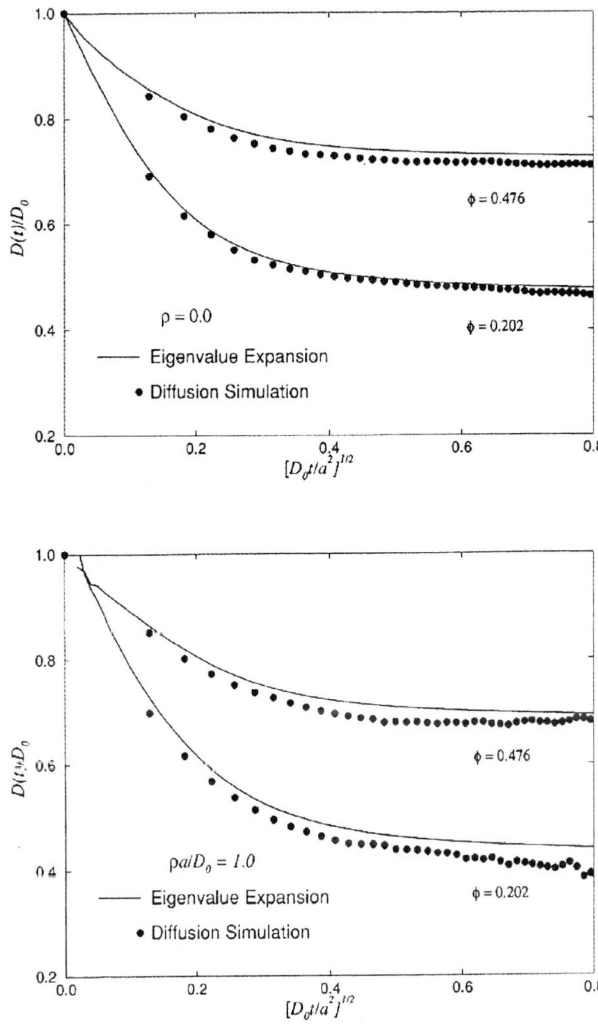

Figure 7.15: Results for $D(t)/D_p$ in the cubic array of spheres microstructure, based on the eigenstate expansion and on random walker simulations, are compared for two values of the relaxation strength ρ at the two porosities, $\phi = 0.4764$ and 0.202. In (a) $\rho = 0$ while in (b) $\rho a/D_0 = 1.0$; in the figure D_p is denoted by D_0 (after Fig. 5 of Ref. [20]).

Appendix A

Diffusion eigenstates and PFGSE amplitude

A serious obstacle in any calculational procedure, for solving the equations either for the spin density $m(\mathbf{r}, t)$ or for the diffusion eigenstates, is the need to satisfy the boundary condition of Eq. (7.3) or Eq. (7.6) at the pore/matrix interface. That is a difficult task for any but the simplest kind of isolated voids microstructure, even when $\rho = 0$. A workable practical method to overcome this difficulty in the latter case is to first extend the partial differential equations (7.2) and (7.5) to the entire system volume V, which includes both the pore space (of volume V_p) and the rock matrix, recasting the problem as a diffusion problem in a composite medium made of two constituents. In order to do that, the diffusion equation must be rewritten in terms of the local chemical potential $\mu(\mathbf{r}, t)$, which is the thermodynamically conjugate variable of the spin density. In contrast with $m(\mathbf{r}, t)$, which even in equilibrium has different values in different constituents, i.e., it is discontinuous across the interface between different constituents, $\mu(\mathbf{r}, t)$ is everywhere a continuous function of $\mathbf{r}$, and has the same constant value everywhere when the system is in equilibrium and there is no net diffusion.

For a system at local equilibrium (i.e., not too far from equilibrium in sufficiently small subvolumes), small changes in $m(\mathbf{r}, t)$ are related to small changes in $\mu(\mathbf{r}, t)$ by means of the local value of the equilibrium thermodynamic derivative $\partial m_{eq}/\partial \mu \equiv m_\mu(\mathbf{r})$, which, like the local value of the diffusion coefficient $D(\mathbf{r})$, has different values in the different constituents. That relation, as well as an expression for $D(\mathbf{r})$, can be written with the help of the characteristic functions of the different constituents—in the present case these functions are $\theta_p(\mathbf{r})$, which equals 1 if $\mathbf{r}$ is in the pore space and 0 otherwise, and $\theta_m(\mathbf{r}) = 1 - \theta_p(\mathbf{r})$, which equals 1 if $\mathbf{r}$ is in the matrix and 0 otherwise:

$$\begin{aligned} \delta m(\mathbf{r},t) &= m_\mu(\mathbf{r})\delta\mu(\mathbf{r},t), \\ m_\mu(\mathbf{r}) &= m_{\mu p}\theta_p(\mathbf{r}) + m_{\mu m}\theta_m(\mathbf{r}), \\ D(\mathbf{r}) &= D_p\theta_p(\mathbf{r}) + D_m\theta_m(\mathbf{r}). \end{aligned}$$

The subscripts p, m denote quantities related to the pore space and the matrix space, respectively. Although both D_m and $m_{\mu m}$ should be taken to vanish, it is sometimes useful, for computational purposes, to allow them to have small positive values. In particular, their relative order of smallness can have a major effect on convergence of the numerical procedure [77].

The case $\rho = 0$

With the help of $D(\mathbf{r})$ and $m_\mu(\mathbf{r})$, which have well defined values everywhere, Eqs. (7.2) and (7.3) can be replaced by a single partial differential equation for $\mu(\mathbf{r},t)$ when $\rho = 0$

$$m_\mu(\mathbf{r})\frac{\partial \mu}{\partial t} = \nabla \cdot [D(\mathbf{r})m_\mu(\mathbf{r})\nabla\mu]. \tag{A.1}$$

The correct boundary condition at the interface between the two constituents, when D_p and D_m have arbitrary values but $\rho = 0$, is that the normal component of the diffusive flux $D(\mathbf{r})m_\mu(\mathbf{r})\nabla\mu$ is continuous there. That condition is *imposed automatically* by requiring that Eq. (A.1) hold everywhere: in the pore space, in the rock matrix, and also at the pore/matrix interface.

The eigenvalue equations (7.5) and (7.6) are now replaced by the single equation (λ represents the eigenvalue and ϕ_λ the eigenfunction)

$$0 = \lambda m_\mu(\mathbf{r})\phi_\lambda(\mathbf{r}) + \nabla \cdot [D(\mathbf{r})m_\mu(\mathbf{r})\nabla\phi_\lambda(\mathbf{r})]. \tag{A.2}$$

Note that, like $\mu(\mathbf{r},t)$, $\phi_\lambda(\mathbf{r})$ is everywhere continuous and is defined throughout the entire system volume V, i.e., in the pore space as well as in the matrix space.

If the microstructure is periodic, the eigenfunctions $\phi_\lambda(\mathbf{r})$ are Bloch functions

$$\phi_\lambda(\mathbf{r}) \to \phi_{n\mathbf{q}}(\mathbf{r})e^{i\mathbf{q}\cdot\mathbf{r}},$$

whose periodic part $\phi_{n\mathbf{q}}(\mathbf{r})$ can be expanded in a Fourier series

$$\phi_{n\mathbf{q}}(\mathbf{r}) = \sum_{\mathbf{g}} \tilde{\phi}_{n\mathbf{q}}(\mathbf{g}) e^{i\mathbf{g}\cdot\mathbf{r}}, \qquad \tilde{\phi}_{n\mathbf{q}}(\mathbf{g}) = \frac{1}{V_a}\int_{V_a} dV\, \phi_{n\mathbf{q}}(\mathbf{r}) e^{-i\mathbf{g}\cdot\mathbf{r}}.$$

Here V_a is the volume of one unit cell, and the sum is over all the vectors $\mathbf{g}$ of the appropriate reciprocal lattice. The eigenvalues now form quasi-continuous bands $\lambda_{n\mathbf{q}}$ and, like the eigenfunctions, they too are characterized by a band index n and a wave vector $\mathbf{q}$, which lies in the first Brillouin zone of reciprocal space. Using the Fourier expansion of $\phi_{n\mathbf{q}}(\mathbf{r})$, the differential equation (A.2) is translated into the following infinite set of linear algebraic equations for the Fourier expansion coefficients $\tilde{\phi}_{n\mathbf{q}}(\mathbf{g})$

$$\begin{aligned} \lambda_{n\mathbf{q}} \sum_{\mathbf{g}'} W_{\mathbf{g}-\mathbf{g}'} \tilde{\phi}_{n\mathbf{q}}(\mathbf{g}') &= D_p \sum_{\mathbf{g}'} \Gamma_{\mathbf{g}\mathbf{g}'}(\mathbf{q}) \tilde{\phi}_{n\mathbf{q}}(\mathbf{g}') \\ \Gamma_{\mathbf{g}\mathbf{g}'}(\mathbf{q}) &= (\mathbf{g}+\mathbf{q})\cdot(\mathbf{g}'+\mathbf{q}) U_{\mathbf{g}-\mathbf{g}'} \\ U_{\mathbf{g}-\mathbf{g}'} &= \delta_{\mathbf{g}\mathbf{g}'} - u\theta^{(m)}_{\mathbf{g}-\mathbf{g}'} \\ W_{\mathbf{g}-\mathbf{g}'} &= \delta_{\mathbf{g}\mathbf{g}'} - w\theta^{(m)}_{\mathbf{g}-\mathbf{g}'}. \end{aligned} \tag{A.3}$$

Here $\theta^{(m)}_{\mathbf{g}}$ is the $\mathbf{g}$-Fourier coefficient of $\theta_m(\mathbf{r})$, and is given by

$$\theta^{(m)}_{\mathbf{g}} = \frac{1}{V_a}\int_{V_a} dV \theta_m(\mathbf{r}) e^{-i\mathbf{g}\cdot\mathbf{r}}, \tag{A.4}$$

while w and u are parameters that characterize the heterogeneous thermodynamic and transport properties of the composite:

$$w \equiv 1 - \frac{m_{\mu m}}{m_{\mu p}}, \qquad u \equiv 1 - \frac{m_{\mu m} D_m}{m_{\mu p} D_p}.$$

Eq. (A.3) is a *generalized matrix eigenvalue problem*, because of the presence of the $\widehat{W}$-matrix on the left-hand-side. In order for this problem to yield to standard numerical methods, it is important to keep $\widehat{W}$ positive definite, hence $m_{\mu m}/m_{\mu p}$ should be kept small but greater than 0 in numerical computations (i.e., w close to but less than 1). No such restriction exists regarding the $\widehat{\Gamma}$-matrix, which is in fact only *positive semi-definite* when $D_m = 0$. If the eigenvectors $\tilde{\phi}_{n\mathbf{q}}(\mathbf{g})$ are normalized as follows ($W(\mathbf{r}) = 1 - w\theta_m(\mathbf{r})$ is the positive periodic function whose Fourier expansion coefficients are $W_{\mathbf{g}}$)

$$\sum_{\mathbf{g},\mathbf{g}'} \tilde{\phi}^*_{m\mathbf{q}}(\mathbf{g}) W_{\mathbf{g}-\mathbf{g}'} \tilde{\phi}_{n\mathbf{q}}(\mathbf{g}') = \frac{1}{V_a}\int_{V_a} dV W(\mathbf{r}) \phi^*_{m\mathbf{q}}(\mathbf{r}) \phi_{n\mathbf{q}}(\mathbf{r}) = \delta_{mn}, \tag{A.5}$$

we get the following expression for the PFGSE amplitude

$$M(\mathbf{k},t) = \frac{1}{\phi}\sum_n e^{-\lambda_{n\mathbf{q}}t}|\widetilde{\psi}_{n\mathbf{q}}(\mathbf{g_k})|^2, \quad \mathbf{q} = \mathbf{k} - \mathbf{g_k}, \tag{A.6}$$

where $\mathbf{g_k} \equiv \mathbf{k} - \mathbf{q}$ is the reciprocal lattice vector that is closest to $\mathbf{k}$, and

$$\widetilde{\psi}_{n\mathbf{q}}(\mathbf{g}) \equiv \frac{1}{V_a}\int_{V_a} dV \theta_p(\mathbf{r})\phi_{n\mathbf{q}}(\mathbf{r})e^{-i\mathbf{g}\cdot\mathbf{r}} = \sum_{\mathbf{g}'} \theta^{(p)}_{\mathbf{g}-\mathbf{g}'}\widetilde{\phi}_{n\mathbf{q}}(\mathbf{g}') \tag{A.7}$$

is the $\mathbf{g}$-Fourier coefficient of $\theta_p(\mathbf{r})\phi_{n\mathbf{q}}(\mathbf{r})$. The latter function is equal to $\phi_{n\mathbf{q}}(\mathbf{r})$ in the pore space and equal to 0 elsewhere. Truncated versions of Eq. (A.3) (i.e., keeping all the $\mathbf{g}$-vectors up to some finite upper limit $|\mathbf{g}| \leq g_{max}$) can be solved numerically, using standard methods for matrix eigenvalue problems. Such calculations are described in Refs. [77, 102].

Fig. A.1 shows results for the lowest bands of eigenvalues $\lambda_{n\mathbf{q}}$ in a particular case of a simple cubic spheres array with $\rho = 0$. Because the eigenstates are found by numerical solution of a truncated version of Eq. (A.3), only a finite portion of these eigenvalues can be obtained accurately, i.e., those that are not too large. The contribution of high eigenstates to macroscopic quantities like $M(\mathbf{k},t)$, $M(t)$, $D(t)$ quickly becomes negligible with increasing t, because they are always weighted by the exponentially decreasing factors $e^{-\lambda_{n\mathbf{q}}t}$. However, for sufficiently short times this method of calculation will always break down.

These calculations lead to results for $D(t)$, using Eq. (7.37), and also for D_e, using the relation [compare with Eq. (7.34)]

$$\lambda_{0\mathbf{q}} \approx D_e\mathbf{q}^2 \text{ for small } \mathbf{q}.$$

Note that, in the correct physical limit $D_m = m_{\mu m} = 0$, w and u are both equal to 1. Thus, $W(\mathbf{r})$ reduces to $\theta_p(\mathbf{r})$, $U_\mathbf{g}$ and $W_\mathbf{g}$ both reduce to

$$\theta_\mathbf{g} = \delta_{\mathbf{g}0} - \theta^{(m)}_\mathbf{g} = \frac{1}{V_a}\int_{V_a} dV\, \theta_p(\mathbf{r})e^{-i\mathbf{g}\cdot\mathbf{r}}, \tag{A.8}$$

and the normalization of $\phi_{n\mathbf{q}}$ reduces from Eq. (A.5) to

$$\frac{1}{V_a}\int_{V_a\cap V_p} dV \phi^*_{m\mathbf{q}}(\mathbf{r})\phi_{n\mathbf{q}}(\mathbf{r}) = \sum_\mathbf{g} \widetilde{\psi}^*_{m\mathbf{q}}(\mathbf{g})\widetilde{\psi}_{n\mathbf{q}}(\mathbf{g}) = \delta_{mn}. \tag{A.9}$$

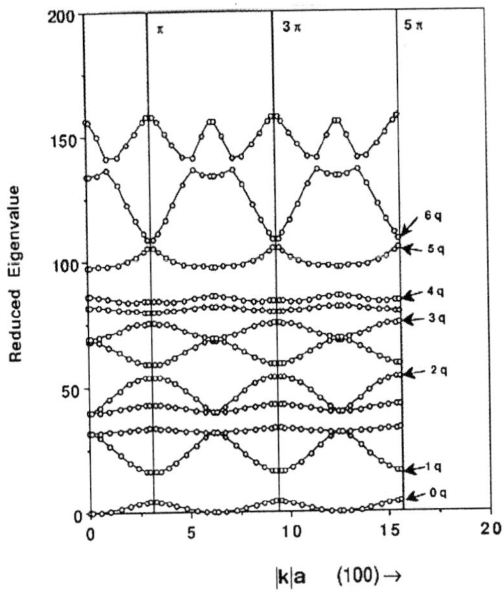

Figure A.1: Results of the 12 lowest reduced (i.e., dimensionless) eigenvalues $\lambda_{n\mathbf{k}}a^2/D_p$, plotted vs. $|\mathbf{k}|a$ along the (100) direction in the one dimensional extended Brillouin zone scheme, for a simple cubic array of identical touching spheres with $\rho = 0$ and ϕ=0.476, where a is the edge length of the unit cell. The bands which are identified as **0q**, **1q**, etc. correspond to the eigenvalues $\lambda_{0\mathbf{q}}$, $\lambda_{1\mathbf{q}}$, etc. The unidentified bands have very small weights and make a negligible contribution to $M(\mathbf{k}, t)$ at all times (after Fig. 2 of Ref. [77]).

The case $\rho \neq 0$

When $\rho \neq 0$, the above approach is not applicable. Instead, one can expand the eigenfunctions in terms of the $\rho = 0$ eigenfunctions for the same microstructure. In the case of a periodic microstructure, both types of eigenfunctions are Bloch functions, and the periodic part of the $\rho \neq 0$ eigenfunctions $\psi_{n\mathbf{q}}(\mathbf{r})$ is expanded in terms of the periodic parts of the $\rho = 0$ eigenfunctions $\phi_{m\mathbf{q}}(\mathbf{r})$:

$$\psi_{n\mathbf{q}} = \sum_m A_m^{(n)}(\mathbf{q})\phi_{m\mathbf{q}}.$$

The functions $\psi_{n\mathbf{q}}(\mathbf{r})$ are only defined inside the pore space, thus their Fourier coefficients are related to $\widetilde{\psi}_{n\mathbf{q}}(\mathbf{g})$ of the $\rho = 0$ problem [see Eq. (A.7)] in the same fashion

$$\widetilde{\psi}_{n\mathbf{q}}(\mathbf{g})\Big|_{\rho\neq 0} = \sum_m A_m^{(n)}(\mathbf{q})\, \widetilde{\psi}_{m\mathbf{q}}(\mathbf{g})\Big|_{\rho=0}.$$

The expansion coefficients $A_m^{(n)}$ are the solution of the following matrix eigenvalue problem

$$(\mu_{n\mathbf{q}} - \lambda_{m\mathbf{q}})A_m^{(n)}(\mathbf{q}) = \rho \sum_p V_{mp}(\mathbf{q})A_p^{(n)}(\mathbf{q}).$$

Here $\mu_{n\mathbf{q}}$ are the eigenvalues for $\rho \neq 0$ and

$$V_{mp}(\mathbf{q}) \equiv \frac{1}{V_a} \oint_{\partial V_p \cap V_a} dS \phi^*_{m\mathbf{q}}(\mathbf{r})\phi_{p\mathbf{q}}(\mathbf{r}) \tag{A.10}$$

are *surface overlap integrals* of two $\rho = 0$ eigenfunctions over the pore/matrix interface ∂V_p in one unit cell, with $\phi_{m\mathbf{q}}(\mathbf{r})$ normalized as in Eq. (A.9).

In practice, the matrix elements $V_{nm}(\mathbf{q})$ are evaluated using the Fourier expansion coefficients of $\theta_p(\mathbf{r})\phi_{m\mathbf{q}}(\mathbf{r})$, i.e., the coefficients $\widetilde{\psi}_{n\mathbf{q}}(\mathbf{g})$ of the $\rho = 0$ problem [see Eq. (A.7)]

$$V_{nm}(\mathbf{q}) = 4 \sum_{\mathbf{g}} \sum_{\mathbf{g}'} \widetilde{\psi}^*_{n\mathbf{q}}(\mathbf{g})\Big|_{\rho=0} K(\mathbf{g}-\mathbf{g}')\, \widetilde{\psi}_{m\mathbf{q}}(\mathbf{g}')\Big|_{\rho=0}, \tag{A.11}$$

where

$$K(\mathbf{g}) \equiv \frac{1}{V_a} \oint_{\partial V_p \cap V_a} dS\, e^{-i\mathbf{g}\cdot\mathbf{r}}. \tag{A.12}$$

A somewhat subtle but important technical point is the factor 4 which appears in Eq. (A.11): It arises from the fact that the function $\theta_p(\mathbf{r})\phi_{n\mathbf{q}}(\mathbf{r})$ jumps from 0 to $\phi_{n\mathbf{q}}(\mathbf{r})$ when one crosses the interface. Therefore, *at the interface* the Fourier series expansion of that discontinuous function actually converges to 1/2 of the nonzero value of $\phi_{n\mathbf{q}}(\mathbf{r})$ there. The values of $K(\mathbf{g})$ in general depend upon details of the microstructure and periodicity, but $K(\mathbf{0})$ has a simple general value, namely

$$K(\mathbf{0}) = \frac{S_p}{V} = \phi \frac{S_p}{V_p}.$$

It is worth noting that the normalized lowest $\rho = 0$ eigenstate, characterized by $n = 0$ and $\mathbf{q} = 0$, is just

$$\phi_{00}(\mathbf{r}) = \left(\frac{1}{\phi}\right)^{1/2}, \tag{A.13}$$

and that the diagonal matrix element of $\widehat{V}$ in that state is

$$V_{00}(\mathbf{0}) = \frac{S_p}{V_p}.$$

If the eigenvectors $A_m^{(n)}(\mathbf{q})$ are normalized to 1, then $M(\mathbf{k}, t)$ is given by an expression like Eq. (A.6), but with $\lambda_{n\mathbf{q}}$ replaced by $\mu_{n\mathbf{q}}$, and with $\widetilde{\psi}_{n\mathbf{q}}(\mathbf{g})$ now representing the Fourier expansion coefficients of the $\rho \neq 0$ eigenfunctions, namely

$$\widetilde{\psi}_{n\mathbf{q}}(\mathbf{g}) \equiv \frac{1}{V_a} \int_{V_p \cap V_a} dV\, \psi_{n\mathbf{q}}(\mathbf{r})\big|_{\rho \neq 0}\, e^{-i\mathbf{g}\cdot\mathbf{r}}.$$

That explains and justifies Eq. (7.63), which can be used both for $\rho = 0$ and $\rho \neq 0$, if $\mu_{n\mathbf{q}}$ and $\widetilde{\psi}_{n\mathbf{q}}(\mathbf{g})$ are chosen appropriately.

Fig. A.2 shows results for the same microstructure as Fig. A.1, but with two nonzero values of ρ. As in the $\rho = 0$ case, here too the high eigenstates will not be calculated reliably because of the matrix truncation. Consequently the calculation of $M(\mathbf{k}, t)$, $M(t)$, $D(t)$ breaks down for sufficiently short t.

As was the case when $\rho = 0$, the low $\mathbf{q}$ eigenvalues of the lowest band provide the value of D_e

$$\mu_{0\mathbf{q}} = \mu_{00} + D_e \mathbf{q}^2 + \mathcal{O}(\mathbf{q}^4),$$

except that μ_{00} is nonzero, and the value of D_e depends on ρ.

The time dependent diffusion coefficient

Using Eq. (7.37), the small $\mathbf{k}$ behavior of $M(\mathbf{k}, t)$ again leads to results for the time dependent diffusion coefficient $D(t)$. However, there is now a somewhat subtle dependence of $D(t)$ on the precise way it is defined. The most natural physical definition of $D(t)$ is based on the mean square displacement of a diffusing particle, and is motivated by the result for a uniform fluid, given by Eq. (7.24):

$$\langle [\mathbf{r}(t) - \mathbf{r}(0)]^2 \rangle \equiv 6D(t)t, \tag{A.14}$$

where the average is restricted to diffusing particles that survive (i.e., spins that did not depolarize) at time t. This average can be calculated as follows:

$$\langle [\mathbf{r}(t) - \mathbf{r}(0)]^2 \rangle = \frac{1}{V_p} \int_{V_p} dV' \frac{\int_{V_p} dV (\mathbf{r} - \mathbf{r}')^2 G(\mathbf{r}, \mathbf{r}'|t)}{\int_{V_p} dV G(\mathbf{r}, \mathbf{r}'|t)}, \tag{A.15}$$

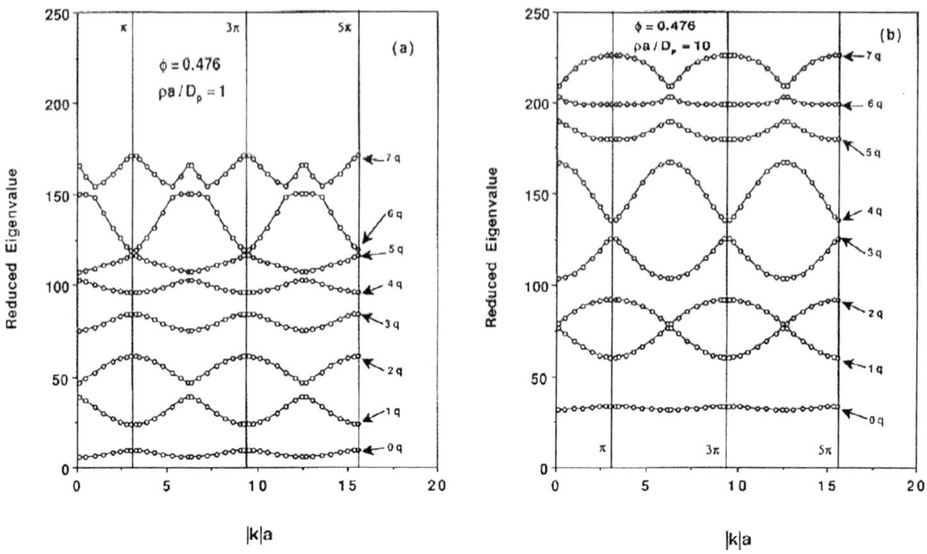

Figure A.2: Results for the 7 lowest reduced (i.e., dimensionless) eigenvalues $\mu_{n\mathbf{k}}a^2/D_p$, plotted vs. $|\mathbf{k}|a$ along the (100) direction in the one dimensional extended zone scheme, for a simple cubic array of identical touching spheres (ϕ=0.4764) (a) for $\rho a/D_p = 1$ (b) for $\rho a/D_p = 10$. The bands which are identified as $0\mathbf{q}$, $1\mathbf{q}$, etc. correspond to the eigenvalues $\mu_{0\mathbf{q}}$, $\mu_{1\mathbf{q}}$, etc. (after Fig. 2 of Ref. [18]).

where the integral $\int_{V_p} dV\, G(\mathbf{r}, \mathbf{r}'|t)$ is the probability of survival of the particle that was originally at $\mathbf{r}'$. Only when the decay of $M(t)$ can be neglected, can this integral be replaced by 1, and only then can we write

$$6D(t)t = -\left.\frac{1}{M(\mathbf{0},t)}\frac{\partial^2 M(\mathbf{k},t)}{\partial\mathbf{k}\cdot\partial\mathbf{k}}\right|_{\mathbf{k}=0}, \tag{A.16}$$

which is equivalent to Eq. (7.37). Although the uniform exponential decay rate $1/T_{2B} + \nu_0$ can be factored out of $M(\mathbf{k}, t)$, the same cannot be done for the ρ dependence of the decay. Thus, when ρ is not small, the $D(t)$ calculated from Eq. (7.37) or Eq. (A.16) will differ from the $D(t)$ of Eq. (A.14). Since Eq. (A.16) is based upon a macroscopically measurable quantity, namely the PFGSE amplitude $M(\mathbf{k}, t)$, it is sometimes preferable to use that equation to define $D(t)$, instead of the more natural intuitive physical definition implied by Eqs. (A.14) and (A.15).

Decay of the total polarization

The $\mathbf{q} = 0$, $n \neq 0$ eigenvalues have non-trivial values for $\rho \neq 0$, as well as for $\rho = 0$. However, when $\rho = 0$, they make no contribution to the decay rate of the total spin polarization $M(t) = M(0)M(\mathbf{k} = 0, t)$, which is constant in time (apart from the constant, microstructure independent decay rate $1/T_{2B} + \nu_0$, which is ignored here). This is due to the fact that the $\mathbf{q} = 0$, $\mathbf{g} = 0$, $n \neq 0$ Fourier expansion coefficients all vanish

$$\widetilde{\psi}_{n\mathbf{0}}(\mathbf{0}) = 0 \quad \text{for} \quad n \neq 0, \quad \text{when} \quad \rho = 0,$$

because $\phi_{n\mathbf{0}}(\mathbf{r})$ is orthogonal to $\phi_{0\mathbf{0}}(\mathbf{r})$, which is a constant.

By contrast, when $\rho \neq 0$, the analogous coefficients do not vanish. Therefore the eigenvalues $\mu_{n\mathbf{0}}$ provide an entire spectrum of non-trivial decay rates for $M(t)$ via

$$M(\mathbf{0}, t) = \frac{1}{\phi} \sum_n e^{-\mu_{n\mathbf{0}} t} |\widetilde{\psi}_{n\mathbf{0}}(\mathbf{0})|^2.$$

Appendix B

Short time asymptotics

The asymptotic treatment of diffusion is based upon the idea that, for sufficiently short times, the effects of the interface are only significant at locations sufficiently close to it that the interface can be approximated by an infinite flat plane. This idea was apparently first expressed in Ref. [166] and was implemented in detail in Ref. [167], where some of the results described in this appendix were first obtained. In that case, the problem reduces to diffusion in a semi-infinite space bounded by that plane, which we take to be $x = 0$.

The diffusion propagator for that problem is a product of three one dimensional diffusion propagators—free or unrestricted propagators G_0 along y and z, e.g.

$$G_0(y, y'|t) = \left(\frac{1}{4\pi D_p t}\right)^{1/2} e^{-\frac{(y-y')^2}{4D_p t}},$$

and one "half space propagator" $G_1(x, x'|t)$ along x. The latter can be written in closed form if $\rho = 0$, and its Laplace transform $\tilde{G}_1(x, x'|s) \equiv \mathcal{L}[G_1(x, x'|t)]$ can be written in closed form when ρ is arbitrary:

$$\begin{aligned}\tilde{G}_1(x, x'|s) &\equiv \mathcal{L}\left[G_1(x, x'|t)\right] \equiv \int_0^\infty dt\, G_1(x, x'|t) e^{-st} \\ &= \frac{1}{2(sD_p)^{1/2}} \left[e^{-(s/D_p)^{1/2}|x-x'|} + \frac{(sD_p)^{1/2} - \rho}{(sD_p)^{1/2} + \rho} e^{-(s/D_p)^{1/2}(x+x')}\right].\end{aligned}$$

The Laplace transform of the relative total polarization $M(t)/M(0) = M(\mathbf{k} = 0, t)$ is then given by

$$\mathcal{L}[M(\mathbf{k} = 0, t)] = \frac{1}{s} - \frac{S_p}{V_p} \frac{D_p^{1/2} \rho}{s^{3/2}[(sD_p)^{1/2} + \rho]}, \tag{B.1}$$

and the transform of the product $M(\mathbf{k}=0,t)D(t)t$, where $D(t)$ is defined by Eq. (7.37) or Eq. (A.16), is given by

$$\mathcal{L}\left[M(\mathbf{k}=0,t)D(t)t\right] = \frac{D_p}{s^2} - \frac{S_p}{V_p}\frac{D_p}{s^2}\left[\frac{2D_p\rho}{(sD_p)^{1/2}[(sD_p)^{1/2}+\rho]} + \right.$$
$$\left. + \frac{D_p}{3(sD_p)^{1/2}}\frac{(sD_p)^{1/2}-\rho}{(sD_p)^{1/2}+\rho} + \frac{D_p\rho}{3[(sD_p)^{1/2}+\rho]^2}\right]. \quad \text{(B.2)}$$

These results are valid as long as $(D_p/s)^{1/2}$ is much less than any finite size feature of the interface, and they can be used to reconstruct $M(t)$, $D(t)$ for times such that $(D_pt)^{1/2}$ is similarly small. This is most conveniently done, in the two limits $\rho t \ll (D_pt)^{1/2}$ and $\rho t \gg (D_pt)^{1/2}$, by first expanding Eqs. (B.1), (B.2) in powers of ρ or $1/\rho$ and then applying the inverse Laplace transformation term by term. This leads to the following results [18]:

$$\frac{M(t)}{M(0)} \cong \begin{cases} 1-\frac{S_p}{V_p}\rho t & \text{for } \rho \ll (D_p/t)^{1/2}, \\ 1-2\frac{S_p}{V_p}\left(\frac{D_pt}{\pi}\right)^{1/2} & \text{for } \rho \gg (D_p/t)^{1/2}, \end{cases} \quad \text{(B.3)}$$

$$\frac{D(t)}{D_p} \cong \begin{cases} 1-\frac{4}{9}\frac{S_p}{V_p}\left(\frac{D_pt}{\pi}\right)^{1/2}+\frac{1}{6}\frac{S_p}{V_p}\rho t & \text{for } \rho \ll (D_p/t)^{1/2}, \\ 1-\frac{2}{9}\frac{S_p}{V_p}\left(\frac{D_pt}{\pi}\right)^{1/2} & \text{for } \rho \gg (D_p/t)^{1/2}. \end{cases} \quad \text{(B.4)}$$

If we use the more physical definition of $D(t)$, namely Eqs. (A.14) and (A.15), we get some slightly different results [15]:

$$\frac{D(t)}{D_p} \cong \begin{cases} 1-\frac{4}{9}\frac{S_p}{V_p}\left(\frac{D_pt}{\pi}\right)^{1/2}+\frac{3}{4}\frac{S_p}{V_p}\rho t & \text{for } \rho \ll (D_p/t)^{1/2}, \\ 1-0.392\frac{2}{9}\frac{S_p}{V_p}\left(\frac{D_pt}{\pi}\right)^{1/2} & \text{for } \rho \gg (D_p/t)^{1/2}. \end{cases} \quad \text{(B.5)}$$

The result in the second line is based on the following integral (erfc(x) is the complementary error function—see Ref. [2])

$$6\int_0^\infty \frac{2\sqrt{\pi}x^2\mathrm{erfc}(x)-xe^{-x^2}}{1-\mathrm{erfc}(x)}dx \cong 6\cdot 0.065352 = 0.392112,$$

which was evaluated numerically using Mathematica 4. Interestingly, the leading order short time correction to D_p is independent of ρ according to both definitions of $D(t)$. At the shortest times, that leading correction is the same for both definitions of $D(t)$—compare the first line in Eqs. (B.4) and (B.5).

Similar considerations, applied to Eq. (7.70), lead to the short time result for $P_s(t)$ of Eq. (7.71)—see Ref. [110].

All the results obtained in this appendix do not include corrections that arise from microgeometrical details other than S_p/V_p, such as the local radii of curvature of the interface. Some of those corrections, which are of higher order in some appropriate characteristic size of the pore space, are discussed and calculated in Ref. [167], while others appear in Appendix C of Ref. [18].

Bibliography

[1] Abragam, A. *Principles of Nuclear Magnetism*, Oxford University Press (1961).

[2] Abramowitz, M. and Stegun, I.A., *Handbook of Mathematical Functions*, Dover Publications (1965).

[3] Akkurt, R., Vinegar, H.J., Tutunjian, P.N., and Guillory, A.J., The Log Analyst, **37**, 33 (1996).

[4] Akkurt, R., Mardon, D., Gardner, J.S., Marshall, D.M., and Solanet, F., SPWLA 39th Annual Logging Symposium, Paper GG, May 26-29 (1998).

[5] Akkurt, R., Marshall, D.M., Eyvazzadeh, R.Y., Gardner, J.S., Mardon, D., Dunn, K.J., SPE Paper 49014, Society of Petroleum Engineers, New Orleans, LA (1998).

[6] Almagor, E. and Belfort, G., J. Colloid Interface Sci., **66**, 146 (1978).

[7] Amott, E., Trans. AIME, **216**, 156 (1959).

[8] Amyx, J.W., Bass, D.M., Jr., Whiting, R.L., *Petroleum Reservoir Engineering*, McGraw-Hill (1988).

[9] Anderson, W.G., *Wettability literature survey*, parts 1 through 6, all in Journal of Petroleum Technology, part 1: vol.38.2, no.11, 1125-1144 (1986); part 2: vol.38.2, no.12, 1246-1262 (1986); part 3: vol.38.2, no.13, 1371-1378 (1986); part 4: vol.39.2, no.10, 1283-1300 (1987); part 5: vol.39.2, no.11, 1453-1468 (1987); and part 6: vol.39.2, no.12, 1605-1622 (1987).

[10] Appel, M., Freeman, J.J., Perkins, R.B., and Hofman, J.P., SPWLA 40th Annual Logging Symposium, Paper FF, Oslo, Norway, May 30-June 3 (1999).

[11] Banavar, J.R. and Schwartz, L.M., Phys. Rev. Lett. **58**, 1411 (1987).

[12] Bendel, P., J. Magn. Reson. **86**, 509 (1990).

[13] Benson, T.B., and MaDonald, P.J., J. Magn. Reson. A **112**, 17 (1995).

[14] Benson, T.B., and MaDonald, P.J., J. Magn. Reson. B **109**, 314 (1995).

[15] Bergman, D.J., unpublished.

[16] Bergman, D.J. and Dunn, K.-J., Phys. Rev. B **45**, 13262 (1992).

[17] Bergman, D.J. and Dunn, K.-J., Phys. Rev. B **50**, 9153 (1994).

[18] Bergman, D.J. and Dunn, K.-J., Phys. Rev. E **51**, 3401 (1995).

[19] Bergman, D.J. and Dunn, K.-J., Phys. Rev. E **52**, 6516 (1995).

[20] Bergman, D.J., Dunn, K.-J., Schwartz, L.M., and Mitra, P.P., Phys. Rev. E **51**, 3393 (1995).

[21] Bergman, D.J., Phys. Rev. E **55**, 4235 (1997).

[22] Bloch, F., Hansen, W.W., and Packard, M, Phys. Rev. **69**, 127 (1946).

[23] Bloch, F., Phys. Rev. **70**, 460 (1946).

[24] Bloembergen, N, Purcell, E.U., and Pound, R.V., Phys. Rev. **73**, 679 (1948).

[25] Bloembergen, N., J. Chem. Phys. **27**, 572 (1957).

[26] Blümich, B., Balibanu, F., Eymael, R., Filip, C., Hailu, K., Klein, M., Kühn, H., Wiesmath, A., and Görke, U., in Proc. 5th Int'l. Conf. on Magnetic Resonance Applications in Porous Media, Bologna, Italy, October 9-11 (2000).

[27] Bobroff, S., Guillot, G., Kassab, G., Rivière, C., and Roussel, J.C., 3rd International Symposium on Evaluation of Reservoir Wettability and Its effect on Oil Recovery, Laramie, Wyoming, September 21-23 (1994).

[28] Borgia, G.C., Brighenti, G., Fantazzini, P., Fanti, G.D., and Mesini, E., SPE Formation Evaluation, 206, Sept. (1992).

[29] Borgia, G.C., Fantazzini, P., Fanti, G., and Mesini, E., J. Petrol. Sci. Eng. **8**, 153 (1992).

[30] Borgia, G.C., Brown, R.J.S., Fantazzini, P., Mesini, E., and Valdrè, G., Il Nuovo Cimento D **14**, 745 (1992).

[31] Borgia, G.C., Brown, R.J.S., and Fantazzini, P., Phys. Rev. E **51**, 2104 (1995).

[32] Borgia, G.C., Brown, R.J.S., and Fantazzini, P., J. Appl. Phys. **79**, 3656 (1996).

[33] Borgia, G.C. and Fantazzini, P., unpublished.

[34] Borgia, G.C., Brown, R.J.S., and Fantazzini, P., J. Mag. Reson., **132**, 65 (1998), ibid. **147**, 273 (2001).

[35] Brown, J.A., Brown, L.F., Jackson, J.A., Milewski, J.V., Travis, B.J., SPE Paper 10813, Socitey of Petroleum Engineers (1982).

[36] Brown, R.J.S., Magn. Reson. Imaging, **14**, 811 (1996).

[37] Brown, R.J.S. and Fatt, I., Petroleum Transaction, AIME, **207**, 262 (1956).

[38] Brown, R.J.S. and Gamson, B.W., J. Pet. Tech., **219**, 201 (1960).

[39] Brown, R.J.S., Phys. Rev. **121**, 1379 (1961).

[40] Brown, R.J.S., Nature, **189**, No. 4762, 387 (1961).

[41] Brown, R.J.S., Bull. American Physical Society (Series II), **1**, 216 (1956).

[42] Brown, R.J.S., Bull. American Physical Society, **3**, 23 (1958).

[43] Brown, R.J.S. and Fantazzini, P., Phys. Rev. B **47**, 14823 (1993).

[44] Brown, R.J.S. and Fantazzini, P., Magn. Reson. Imaging **12**, 175 (1994).

[45] Brownstein, K.R. and Tarr, C.E., Phys. Rev. A **19**, 2446 (1979).

[46] Bryar, T.R., Daughney, C.J., and Knight, R.J., J. Magn. Reson. **142**, 74 (2000).

[47] Burnett, L.J. and Jackson, J.A., J. Magn. Reson. **40**, 406 (1980).

[48] Butler, J.P., Reeds, J.A., and Dawson, S.V., SIAM J. Numer. Anal. **18**, 381 (1981).

[49] Callaghan, P.L., *Principles of Nuclear Magnetic Resonance Microscopy*, Oxford University Press Inc., New York (1993).

[50] Callaghan, P.T., Coy, A., MacGowan, D., Packer, K.J., and Zelaya, F.O., Nature **351**, 467 (1991).

[51] Callaghan, P.T., MacGowan, D., Packer, K.J., and Zelaya, F.O., Mag. Res. Imaging **9**, 663 (1991).

[52] Callaghan, P.T., Coy, A., Halpin, T.P.J., MacGowan, D., Packer, K.J., and Zelaya, F.O., J. Chem. Phys. **97**, 651 (1992).

[53] Carman, P.C., *Flow of Gases Through Porous Media*, Acad. Press, Inc., New York (1956).

[54] Carr, H.Y. and Purcell, E.M., Phys. Rev. **94**, 630 (1954).

[55] Chandler, R.N., Drack, E.O., Miller, M.N., and Prammer, M.G., SPE Paper 28365, Society of Petroleum Engineers, New Orleans, LA (1994).

[56] Chang, D., Vinegar, H.J., Morris, C.E., and Straley, C., SPWLA 35th Annual Symposium, Paper A (1994).

[57] Chen, R., Stallworth, P.E., Greenbaum, S.G., and Kleinberg, R.L., J. Mag. Reson., **A110**, 77 (1994).

[58] Chen, S., Olima, O., Gamin, H., Georgi, D.T., and Minetto, C., SPE Paper 49009, Society of Petroleum Engineers (1998).

[59] Chen, S., Georgi, D.T., Withjack, E.M., Minetto, C., Olima, O., and Gamin, H., Petrophysics, **41**, 33 (2000).

[60] Chen, S., Georgi, D., Liu, C., and Thern, H.F., SPWLA 41st Annual Logging Symposium, Paper W, June 4-7 (2000).

[61] Chitale, D.V., Gardner, J., and Sigal, R., SPWLA 41st Annual Logging Symposium, Paper X, Dallas, TX, June 4-7 (2000).

[62] Clow, H., Percival, W.S., and Walters, P.E., U.S. Patent 4,629,986 (1986).

[63] Coates, G.R. and Dumanoir, J.L., The Log Analyst, **15**, No.1 (1974).

[64] Coates, G.R., Marschall, D., Mardon, D., and Galford, J., SPWLA 38th Annual Logging Symposium, Paper QQ, June 15-18 (1997).

[65] Cook, J.C., "Electromagnetic Resonance Borehole Assay Logging," U.S. Bureau of Mines Report OFR-69-78, U.S. Department of Commerce, National Technical Information Service, Accession No. PB 283734.

[66] Coolidge, J.E., Ritchie, I., Gamson, B.W., and Shane, L., Paper 11, Third Annual Logging Symposium of Society of Professional Well Log Analysts, Houston, TX (1962).

[67] Cooper, R.K. and Jackson, J.A., J. Magn. Reson. **41**, 400 (1980).

[68] Counsell, C., Levitt, M.H., and Ernst, R.R., J. Magn. Reson., **63**, 133 (1985).

[69] Craig, F.F., *The Reservoir Engineering Aspects of Waterflooding*, Monograph Volume 3 of the Henry L. Doherty Series, Society of Petroleum Engineers (1971).

[70] Crowe, M.B., Jasek, D.E., Van Dalen, S.C., LaTorraca, G.A., Dunn, K.J., Donovan, M.T., SPWLA 38th Annual Logging Symposium, Paper JJ, June 15-18 (1997).

[71] de Swiet, T.M., Tomaselli, M., Hürlimann, M.D., and Pines, A., J. Mag. Reson., **133**, 385 (1998).

[72] D'Orazio, F., Tarczon, J.C., Halperin, W.P., Eguchi, K., and Mizusaki, T., J. Appl. Phys. **65**, 742 (1989).

[73] D'Orazio, F., Bhattacharja, S., Halperin, W.P., Eguchi, K., and Mizusaki, T., Phys. Rev. **B 42**, 9810 (1990).

[74] Doughty, D.A. and Tomutsu, L., *Investigation of wettability by NMR microscopy and spin-relaxation*, NIPER-720, prepared for U. S. Department of Energy (1993).

[75] Dullien, F.A.L., *Porous Media*, Academic, New York (1979).

[76] Dunn, K.-J., LaTorraca, G.A. Warner, J.L., and Bergman, D.J., SPE Paper 28367, Society of Petroleum Engineers, New Orleans, LA (1994).

[77] Dunn, K.-J. and Bergman, D.J., J. Chem. Phys. **102**, 3041 (1995).

[78] Dunn, K.-J., LaTorraca, G.A., and Bergman, D.J., Geophysics, **64**, 470 (1999).

[79] Dunn, K.-J. and LaTorraca, G.A., J. Magn. Reson., **140**, 153 (1999).

[80] Dunn, K.-J., Magn. Reson. Imaging, **19**, 439 (2000).

[81] Fadeev, D.K., Trudy Mat. Inst. Akad. Nauk SSSR **53**, 387 (1959).

[82] Flaum, C., Kleinberg, R.L., and Hürlimann, M.D., SPWLA 37th Annual Logging Symposium, Paper L, June 16-19 (1996).

[83] Flaum, C., Kleinberg, R.L., and Bedford, J., SPWLA 39th Annual Logging Symposium, Paper UU, May 26-29 (1998).

[84] Flaum, C., Guru, U., and Banerjee, S., SPWLA 41st Annual Logging Symposium, Paper HHH, June 4-7 (2000).

[85] Foley, I., Farooqui, S.A., and Kleinberg, R.L., J. Magn. Reson. **A 123**, 95 (1996).

[86] Fordham, E.J., Flaum, M., Ramakrishnan, T.S., Kenyon, W.E., and Allen, D., U.S. Patent 6,115,671 (2000).

[87] Freedman R. and Morriss, C.E., SPE Paper 30560, Society of Petroleum Engineers (1995).

[88] Freedman, R., U.S. Patents 5,291,137 (1994) and 5,381,092 (1995).

[89] Freedman, R., Cao Minh, C., Gubelin, G., McGinness, T., Terry, B., Freeman, J., and Rawlence, D., SPWLA 39th Annual Logging Symposium, Paper II, May 26-29 (1998).

[90] Freedman, R., Sezginer, A., Flaum, M., Matteson, A., Lo, S., and Hirasaki, G.J., SPE Paper 63214, Society of Petroleum Engineers, Dallas, TX (2000).

[91] Gallegos, D.P., Smith, D.M., and Brinker, C.J., J. Colloid Interface Sci. **124**, 186 (1968).

[92] Gaymard, R. and Poupon, A., The Log Analyst, **9**, 3 (1968).

[93] Gerritsma, C.J., Oosting, P.H., and Trappeniers, N.J., Physica, **51**, 381 (1971).

[94] Goelman, G. and Prammer, M.G., J. Magn. Reson., **113**, 11 (1995).

[95] Guyer, R.A., Phys. Rev. B **48**, 6007 (1993).

[96] Hahn, E.L., Phys. Rev. **80**, 580 (1950).

[97] Halperin, W.P., D'Orazio, F., Bhattacharija, S., and Tarczon, J.C., in *Molecular Dynamics in Restricted Geometries*, edited by Klafter, J. and Drake, J.M. (Wiley, New York, 1989), pp. 311-350.

[98] Hassoun, T.H., Zainalabedin, K., Cao Minh, C., SPE Paper 37770, Society of Petroleum Engineers, Richardson, TX (1997).

[99] Harris, K.R., Physica **94A**, 448 (1978).

[100] Herrick, R.C., Couturie, S.H., and Best, D.L., SPE Paper 8361, SPE Annual Conference, Las Vegas, NV, Sept. 23-26 (1979).

[101] Hinedi, Z.R., Chang, A.C., Anderson, M.A., and Borchardt, D.B., Water Resour. Res. **33**, 2697 (1997).

[102] Hizi, U. and Bergman, D.J., J. Appl. Phys. **87**, 1704 (2000).

[103] Horkowitz, J.P., Hartman, D.E., Clerke, E.A., Vinegar, H.J., and Coates, G.R., SPWLA 36th Annual Logging Symposium, Paper Q, June 26-29 (1995), also in The Log Analyst, **38**, 73 (1997).

[104] Howard, J.J., Kenyon, W.E., and Straley, C., SPE Paper 20600, Society of Petroleum Engineers, New Orleans, LA, Sept. 23-26 (1990). or SPE Formation Evaluation, 193 September (1993).

[105] Howard, J.J. and Spinler, E.A., SPE Paper 26471, Society of Petroleum Engineers, Houston, TX, Oct. 3-6 (1993).

[106] Howard, J.J., Mag. Reson. Imaging, **16**, 529 (1998).

[107] Hsu, W.-F., Li, X., and Flumerfelt, R.W., SPE Paper 24761, Society of Petroleum Engineers (1992).

[108] Hürlimann, M.D., Latour, L.L., and Sotak, C.H., Magn. Reson. Imaging, **12**, 325 (1994).

[109] Hürlimann, M.D., J. Magn. Reson., **131**, 232 (1998).

[110] Hürlimann, M.D., Schwartz, L.M., and Sen, P.N., Phys. Rev. B **51**, 14936 (1995).

[111] Hürlimann, M.D. and Griffin, D.D., J. Magn. Reson., **143**, 120 (2000).

[112] Jackson, J.A., Burnett, L.J., and Harmon, J.F., J. Magn. Reson. **41**, 411 (1980).

[113] Johnson, D.L., Koplik, J., and Schwartz, L.M., Phys. Rev. Lett. **57** 2564 (1986).

[114] Karlicek, R.F. and Lowe, I.J., J. Magn. Reson., **37**, 75 (1980).

[115] Katz, A.J. and Thompson, A.H., Phys. Rev. B **34**, 8179 (1986).

[116] Kenyon, W.E., Day, P.I., Straley, C., and Willemsen, J.F., SPE Paper 15643, Society of Petroleum Engineers, New Orleans, LA (1986).

[117] Kenyon, W.E., Howard, J.J., Sezginer, A., Straley, C., Matteson, A., Horkowitz, K., and Ehrlich, R., SPWLA 30th Annual Logging Symposium, Paper LL (1989).

[118] Kenyon, W.E., Nuclear Geophysics, **6**, 153 (1992).

[119] Kenyon, W.E., The Log Analyst, **38**, 21 (1997).

[120] Kenyon, W.E., Straley, C., Sen, P.N., Herron, M., Matteson, A., Takezaki, H., and Petricola, M.J., SPE Paper 29886, Society of Petroleum Engineers, Richardson, TX (1995).

[121] Kenyon, W.E., Day, P.I., Straley, C., and Willemsen, J.F., *SPE Formation Evaluation*, **3** 622 (1988).

[122] Kenyon, W.E., and Kolleeny, J.A., J. Colloid Interface Sci. **170**, 502 (1995).

[123] Kleinberg, R.L., *Encyclopedia of Nuclear Magnetic Resonance* Vol. 8, pp. 4960-4969, John Wiley & Sons, Chichester (1996).

[124] Kleinberg, R.L., Sezginer, A., Griffin, D.D., and Fukuhara, M., J. Magn. Reson. **97**, 466 (1992).

[125] Kleinberg, R.L., Magnetic Resonance Imaging, **14**, 761 (1996).

[126] Kleinberg, R.L., Farooqui, S.A., and Horsfield, M.A., J. Colloid Interface Sci., **158**, 195 (1993).

[127] Kleinberg, R.L., Straley, C., Kenyon, W.E., Akkurt, R., and Farooqui, S.A., SPE Paper 26470, Society of Petroleum Engineer (1993).

[128] Kleinberg, R.L. and Vinegar, H.J., The Log Analyst, Nov.-Dec., 20 (1996).

[129] Korringa, J., Seevers, D.O., and Torrey, H.C., Phys. Rev. **127**, 1143 (1962).

[130] Kovscek, A.R., Wong, H., and Radke, C.J., American Institute of Chemical Engineers Journal, vol. 39, no. 6, 1072 (1993).

[131] Lanczos, C., *Linear Differential Operators* Van Nostrand (London), Chapter 3 (1961).

[132] LaTorraca, G.A., Dunn, K-J., and Brown, R.J.S., Society of Core Analysts, Paper 9312 (1993).

[133] LaTorraca, G.A., Dunn, K.-J. and Bergman, D.J., SPWLA 36th Annual Logging Symposium, Paper JJ, Paris, France, June 26-29 (1995).

[134] LaTorraca, G.A., Dunn, K.J., Webber, P.R., and Carlson, R.M., Mag. Res. Imaging, **16**, 659 (1998).

[135] LaTorraca, G.A., Stonard, S.W., Webber, P.R., Carlson, R.M., and Dunn, K.J., SPWLA 40th Annual Logging Symposium, Paper PPP, Oslo, Norway, May 30-June 3 (1999).

[136] LaTorraca, G.A., Webber, P.R., and Dunn, K.J., unpublished (1998).

[137] LaTorraca, G.A., Webber, P.R., Dunn, K.J., Carlson, R.M., unpublished (1997).

[138] Latour, L.L., Kleinberg, R.L., and Sezginer, A., J. Colloid Interface Sci. **150**, 535 (1992).

[139] Latour, L.L., Kleinberg, R.L., Mitra, P.P., and Sotak, C.H., J. Magn. Reson. **112**, 83 (1995).

[140] Lawson, C.L. and Hanson, R.J., 1974, *Solving Least Squares Problems* Prentice Hall, New Jersey (1974).

[141] Le Doussal, P. and Sen, P.N., Phys. Rev. B **46**, 3465 (1992).

[142] Liaw, H.K., Kulkarni, R., Chen, S., and Watson, A.T., AIChE J. **42**, 538 (1996).

[143] Lo, S-W., Hirasaki, G.J., Kobayashi, R., and House, W.V., The Log Analyst, Nov.-Dec. 43 (1998).

[144] Lo, S-W., Hirasaki, G.J., House, W.V., and Kobayashi, R., SPE Paper 63217, Society of Petroleum Engineers, Dallas, TX (2000).

[145] Looyestijn, W.J., SPWLA 37th Annual Logging Symposium, Paper SS, June 16-19 (1996).

[179] Prammer, M.G., Drack, E., Goodman, G., Masak, P., Menger, S., Morys, M., Zannoni, S., Suddarth, B., and Dudley, J., SPE Paper 62981 (2000).

[180] Press, W.H., Flannery, B.P., Teukolsky, S.A., and Vettering, W.T., *Numerical Recipes*, Cambridge University Press (1986).

[181] Purcell, E.U., Torrey, H.C., and Pound, R.V., Phys. Rev. **69**, 37 (1946).

[182] Rajan, S. and Lalita, K., J. Magn. Reson. **16**, 115 (1974).

[183] Ramakrishnan, T.S., Schwartz, L.M., Fordham, E.J., Kenyon, W.E., and Wilkinson, D.J., SPWLA 39th Annual Logging Symposium, Paper SS (1998).

[184] Ramakrishnan, T.S., Fordham, E.J., Venkataramanan, L., Flaum, M., and Schwartz, L.M., SPWLA 40th Annual Logging Symposium, Paper MMM (1999).

[185] Reiderman, A., Itskovich, G., Krugliak, Z., and Beard, D.R., in Proc. 5th Int'l. Conf. on Magnetic Resonance Applications in Porous Media, Bologna, Italy, October 9-11 (2000).

[186] Roberts, S.P., Macdonald, P.J., and Tritchard, T., J. Magn. Reson. **A 116**, 189 (1995).

[187] Robertson, B., Phys. Rev. **151**, 273 (1966).

[188] Robinson, J.D., Loren, J.D., and Higdon, W.T., US Patent No. 3,657,730, April (1972).

[189] Robinson, J.D., Loren, J.D., Vagnar, E.A., and Hartman, D.E., J. Petro. Tech. **26**, 226 (1974).

[190] Salathiel, R.A., J. Pet. Tech., vol. 25.2, 1216 (1973).

[191] Saraf, D.N., Kumar, J., and Fatt, I., Indian Journal of Technology, **8**, 125 (1970).

[192] Schwartz, L.M., Hürlimann, M.D., Dunn, K.-J., Mitra, P.P., and Bergman, D.J., Phys. Rev. E **55**, 4225 (1997).

[193] Schwartz, L.M., private communication.

[194] Seevers, D.O., SPWLA 7th Annual Logging Symposium, Paper L, (1966); SPE Paper 3847 (1972).

[195] Sen, P.N., Straley, C., Kenyon, W.E., and Whittingham, M.S., Geophysics, **55**, 61 (1990).

[196] Sen, P.N., Schwartz, L.M., Mitra, P.P., and Halperin, B.I., Phys. Rev. B **49**, 215 (1994).

[197] Sengers, J.V. and Watson, J.T.R., J. Phys. Chem. Ref. Data., **15**, 1291 (1986).

[198] Setser, G.G., and Williams, M.R., SPE Paper 14276, Society of Petroleum Engineers, Richardson, TX (1985).

[199] Sezginer, A., Cao Minh, C., Heaton, N., Herron, M., Freedman, R., and Van Dort, G., SPWLA 40th Annual Logging Symposium, Paper NNN, Oslo, Norway, May 30-June 3 (1999).

[200] Shtrikman, S. and Taicher, Z., U.S. Patent 4,710,713 (1987).

[201] Slichter, C.P., *Principles of Magnetic Resonance*, Springer-Verlag (1978).

[202] Slijkerman, W.J., and Hofman, J.P., Magn. Reson. Imaging, **16**, 541 (1998).

[203] Slijkerman, W.F.J., et al. SPE Paper 56768, Society of Petroleum Engineers, Houston, TX (1999).

[204] Spindler, E.A., SPE Paper 38733, Society of Petroleum Engineers (1997).

[205] Stejskal, E.O. and Tanner, J.E., J. Chem. Phys. **42**, 288 (1965).

[206] Stonard, W.S., LaTorraca, G.A., and Dunn, K.J., unpublished (1997).

[207] Straley, C., Matteson, A., Feng, S., Schwartz, L.M., Kenyon, W.E., and Banavar, J.R. Appl. Phys. Lett. **51**, 1146 (1987).

[208] Straley, C., Rossini, D., Vinegar, H.J., Tutunjian, P.N., and Morriss, C.E., SCA Paper 9494, Society of Core Analysts (1995), or The Log Analyst, **38**, 84 (1997).

[209] Tanner, J.E. and Stejskal, E.O., J. Chem. Phys. **49**, 1768 (1968).

[210] Tanner, J.E., J. Chem. Phys. **52**, 2523 (1970).

[211] Tikhonov, A.N., Soviet Math. Dokl. **4**, 1624 (1963).

[212] Timur, A., SPWLA 9th Annual Logging Symposium, Paper K, New Orleans, LA., June 23-26 (1968).

[213] Timur, A., J. Pet. Tech. 775, June (1969).

[214] Timur, A., The Log Analyst, **9**, No. 4, 8 (1968).

[215] Torrey, H.C., Phys. Rev., **104**, 563 (1956).

[216] Torrey, H.C., Korringa, J., Seevers, D.O., and Uebersfeld, J., Phys. Rev. Letters, **3**, 418 (1959).

[217] Tul'bovich, B.I., Slisarenko, F.A., Klyaev, V.I., Sorokin, S.I., Izv. Vyssh. Ucheb. Zavel., Khim. Tekknol. **9**, 735 (1966).

[218] Turchin, V.F., Kozlov, V.P., and Malkevich, M.S., Soviet Physics Uspekhi, a translation of Uspekhi Fizicheskikh Nauk, **13**, 681 (1971).

[219] Twomay, S., J. Assoc. Comput. Math. **10**, 97 (1963).

[220] Vinegar, H.J., SPWLA short course notes on NMR, Society of Professional Well Log Analysts, Houston, TX (1995).

[221] Wahba, G., SIAM J. Numer. Anal. **14**, 651 (1977).

[222] Walsh, J.B. and Brace, W.F., J. Geophy. Res. **89**, 9425 (1984).

[223] Wayne, R.C. and Cotts, R.M., Phys. Rev. **151**, 264 (1966).

[224] Williams, C. and Fung, B.M., J. Magn. Reson., **50**, 71 (1982).

[225] Wilson, D.M. and LaTorraca, G.A., Paper 9923, International Symposium of the Society of Core Analysts, Golden, CO (1999).

[226] Woessner, D.E., Concepts in Magn. Reson. **13**, 77 (2001).

[227] Vold, R.L., Vold, R.R., and Simon, H.E., J. Magn. Reson., **11**, 283 (1973).

[228] Xiao, L., Du, Y., and Ye, C., J. Colloid and Interface Sci., **164**, 495 (1994).

[229] Zega, J.A., House, W.V., and Kobayashi, R., Ind. Eng. Chem. Res., **29**, 909 (1990).

[230] Zhang, G.Q., Lo, S.-W., Huang, C.C., Hirasaki, G.J., Kobayashi, R., and House, W.V., SPWLA 39th Annual Logging Symposium, Paper FF, Keystone, CO., May 26-29 (1998).

[231] Zhang, G.Q., Hirasaki, G.J., and House, W.V., SPWLA 41st Annual Logging Symposium, Paper AA, Dalla, TX., June 4-7 (2000).

[232] Zhang, G.Q., Huang, C., and Hirasaki, G.J., Petrophysics, **41**, 223 (2000).

GLOSSARY

API Gravity A scale (API or American Petroleum Institute scale) commonly used in the petroleum industry to characterize the quality of a crude oil. It is defined as follows

$$\text{API gravity} \equiv \frac{141.5}{\rho} - 131.5,$$

where ρ is the density of the oil in gm/cm^3.

BRD The Butler-Reeds-Dawson method [48] for solving the Fredholm equation of the first kind with a solution commensurate with the error bar of the data.

Bulk Volume Irreducible The part of the porous earth formation occupied by immobile, capillary bound water as well as the clay bound water. It is also called the irreducible water.

BVI See Bulk Volume Irreducible.

Carr-Purcell-Meiboom-Gill Pulse Sequence A pulse sequence used to measure T_2 relaxation time. It was first suggested by Carr and Purcell [54] and later modified by Meiboom and Gill [161]. The pulse sequence consists of an initial $\pi/2$ pulse and successive π pulses, where the successive π pulses are separated by 2τ and are excited at time τ, 3τ, 5τ,... The spin echoes occur at time 2τ, 4τ, 6τ,..., etc..

CMR™ [1] The Schlumberger Combinable Magnetic Resonance logging tool.

CPMG See Carr-Purcell-Meiboom-Gill Pulse Sequence.

CVM Constituent Viscosity Model, a phenomenological model suggested by Freedman et al. [90] to link the relationships between relaxation time vs viscosity and diffusion coefficient vs viscosity for crude oils.

Differential Spectrum Method A hydrocarbon typing method which uses different repeat delays and the large T_1 contrasts between gas, oil, and water to detect the presence and estimate the quantity of hydrocarbons.

---

[1] Trademark of Schlumberger Corporation

DSM See Differential Spectrum Method.

EDM See Enhanced Diffusion Method.

Enhanced Diffusion Method A hydrocarbon typing method which uses a single large TE and two different wait times, one allows full polarization and another allows partial polarization of the hydrogen nuclei.

FFI See Free Fluid Index.

FID See Free Induction Decay.

Formation Factor The ratio of the conductivity of a fluid to that of a porous medium saturated by the same fluid, where the conductivity of the solid matrix is assumed to be zero.

Free Fluid Index The part of the porous earth formation occupied by mobile water or producible fluids.

Free Induction Decay It is the signal detected in the rf coil immediately after the a single rf pulse, usually a 90° pulse.

GOR See Gas Oil Ratio.

Gas Oil Ratio The Gas Oil Ratio (GOR) is defined as

$$\mathrm{GOR} = \frac{\text{cubic meters of methane at standard condition}}{\text{cubic meters of stock tank oil}},$$

where the standard condition is 60°F and 1 atmospheric pressure.

Gyromagnetic Ratio The ratio of magnetic moment μ to the spin angular momentum $\mathbf{J}$, denoted by γ. For protons, $\gamma_p = 2.675197 \times 10^8$ rad s^{-1} Tesla^{-1} (or $\times 10^4$ rad s^{-1}Gauss^{-1}), or $\gamma_p/2\pi = 4.257707 \times 10^7$ Hz/Tesla (or $\times 10^3$ Hz/Gauss).

HI See Hydrogen Index.

Hydrocarbon Typing Methods or procedures for characterizing the hydrocarbon types, i.e., gas, or oil (high or low viscosity).

Hydrogen Index It is defined as the amount of hydrogen per unit volume of a fluid divided by the amount of hydrogen per unit volume of pure water in standard conditions.

Irreducible Water Saturation The portion of the water in the porous formation that cannot be produced under a certain capillary pressure.

Larmor Frequency The frequency at which nuclear spins precess about the applied static magnetic field. It is given by $f = \gamma H/(2\pi)$, where γ is the gyromagnetic ratio and H is the applied magnetic field.

Longitudinal Relaxation Time Also called Spin-Lattice Relaxation Time. It is a time constant, denoted as T_1, which characterizes the relaxation process of energy transfer between the magnetic energy of the spin system and the thermal energy of its surroundings. If the static magnetic field is along the z direction and the equilibrium magnetization is M_0, Bloch introduced the assumption that the change of M_z with time can be described by the following equations

$$\frac{dM_z}{dt} = \frac{M_0 - M_z}{T_1}.$$

Magnetization The vector sum of magnetic moments per unit volume.

MAS Magic Angle Spinning, a technique routinely used in solid state NMR to average out chemical shift anisotropy and internuclear dipolar interactions.

MRIL™ [2] The Halliburton/Numar's Magnetic Resonance Imaging Logging tool.

MRF Magnetic Resonance Fluid characterization method. It is a method developed by Schlumberger for characterizing formation fluids using information obtained from multiple NMR logging measurements.

NML™ [3] The Nuclear Magnetism Logging tools, operated by the free precession of spins in the earth's magnetic field. Operation of this tool required special doping of the borehole fluid. It has become obsolete after the introduction of modern pulsed NMR logging tools in the 1990's.

NMR See Nuclear Magnetic Resonance.

---

[2] Trademark of Halliburton/Numar Corporation

[3] Trademark of Schlumberger Corporation

Nuclear Magnetic Resonance A physical phenomenon discovered by Bloch and Purcell independently in 1946, for which both of them were awarded Nobel Prize in 1952. It is a phenomenon found in systems of nuclei that possess both magnetic moments and angular momentum. At resonance condition, radio frequency excitations, tuned to the characteristic frequency of the magnetic system, i.e., the Larmor frequency, are used to stimulate the magnetic nuclei which absorb and re-emit energy through interactions with other nuclei undergoing thermal motions.

OBM Oil Based Mud.

OBMF Oil Based Mud Filtrate.

PAPs See Phase-Alternated Pairs.

PFGSE See Pulsed Field Gradient Spin Echo.

Phase-Alternated Pairs Two CPMG echo trains, the second one is acquired with a phase which differs from that of the first one by 180°. The two echo trains are combined to remove the effects of ringing and baseline offset.

Porosity Unit Porosity of a rock is usually expressed in volume fractions, e.g., 0.25, or in percentage, i.e., 25%, which means 25% of the bulk volume is pore space. Porosity unit is a unit commonly used in the petroleum industry, which is numerically the same as the percentage, e.g., 25 p.u. represents a porosity of 0.25 in volume fraction or 25% in percentage.

p.u. See porosity unit.

Pulsed Field Gradient Spin Echo A pulsed field gradient technique frequently used for determining the self diffusion coefficient of fluids and for studying the physical properties of porous media. It consists of field gradient pulses separated by various lengths of diffusion time.

RD See Repeat Delay.

Repeat Delay Between successive π pulse excitations in a CPMG experiment, a time period is inserted to ensure all spins fully return to their thermal equilibrium state along the direction of the static magnetic field and are ready for the next excitation. This time period is referred to as "Repeat Delay", "Wait Time", or "Repeat Time" (and abbreviated as RD, WT, TW, or TR in the literature).

Restricted Diffusion The diffusion of particles, which occurs within the pore space of a porous medium, is restricted by the presence of pore walls, and has a particle density profile different from that in an infinite medium, hence the term, restricted diffusion.

Return To Origin Probability The probability of a diffusing particle to return to its original starting point, i.e., the origin, after a certain length of diffusion time.

rf pulse Radio frequency pulse, which has a carrier frequency equal to the Larmor frequency of the spins and a width (or duration) determined by the experimental condition.

RTOP See Return To Origin Probability.

Shift Spectrum Method A method for hydrocarbon typing that uses different TEs to differentiate oil and water in the T_2 distributions.

Singular Value Decomposition A method in linear algebra for solving simultaneous linear equations.

Spin Echo A refocusing of the dephased spins due to field inhomogeneity in the plane perpendicular to the applied static magnetic field, first discovered by Hahn [96] in 1950.

Spin-Lattice Relaxation Time See Longitudinal Relaxation Time.

Spin-Spin Relaxation time See Transverse Relaxation Time.

SSM See Shift Spectrum Method.

Stimulated Echo Stimulated echo is the echo occurred at $\tau_1+\tau_2$ after three 90° pulses. It was discussed in Hahn's paper on spin echoes [96]. There are additional echoes at $2\tau_2-2\tau_1$, $2\tau_2-\tau_1$, and $2\tau_2$ when $2\tau_1 < \tau_2 < T_2$. For $\tau_1 < \tau_2 < 2\tau_1$, the signal at $2\tau_2 - 2\tau_1$ is missing, but the other remain. Because of the large field inhomogeneities in the probed zone of the logging tools, stimulated echoes occurred in CPMG echo trains in the logging measurements, and corrections are required to remove this effect.

Surface Relaxivity A quantity that measures the strength of the solid/fluid interface of a porous medium on how fast it can relax the spins in the pore fluid. It has a unit of length/time.

SVD See Singular Value Decomposition method.

T_1 See Longitudinal Relaxation Time.

T_2 See Transverse Relaxation Time.

T_2^* The decay time constant of a Free Induction Decay.

$T_{2,LM}$ The logarithmic average of the T_2 relaxation times.

TDA See Time Domain Analysis.

Time Domain Analysis Subtract the echo train of short wait time from the echo train of long wait time and invert the difference of the two echo trains to obtain a T_2 distribution which reflects the signal due to the difference in wait times. This analysis is called Time Domain Analysis.

TE Time between successive echoes in a CPMG echo train. It is equal to 2τ, where τ is the time between the initial 90° rf pulse and the subsequent 180° pulse.

Transverse Relaxation Time Also called Spin-Spin Relaxation Time. It is a time constant, denoted as T_2, which characterizes the spin relaxation process in the plane perpendicular to the applied static magnetic field direction. If the static magnetic field is along the z direction, the decay of the magnetization in the xy plane (i.e., M_x and M_y) is described by the following Bloch equations

$$\frac{dM_x}{dt} = -\frac{M_x}{T_2}, \qquad \frac{dM_y}{dt} = -\frac{M_y}{T_2}.$$

TR See Repeat Delay.

TW See Repeat Delay.

UPEN Uniform Penalty method, an inversion method for NMR data processing suggested by Brown and his colleagues [34] which provides an uniform penalty along the relaxation time axis.

WT See Wait Time.

Wait Time See Repeat Delay.

Wettability It is defined as the tendency of one fluid to spread on or adhere to a solid surface in the presence of other immiscible fluids.

ϕ Notation used in petrophysics to denote the volume fraction of the pore space. $\phi \equiv V_p/V$, where V_p and V are the pore volume and total volume, respectively.

Index